Multiple Choice Questions (MCQ)

in

Food Technology

Multiple Choice Questions (MCQ) in Food Technology

by
P.M. Kotecha
U.D. Chavan

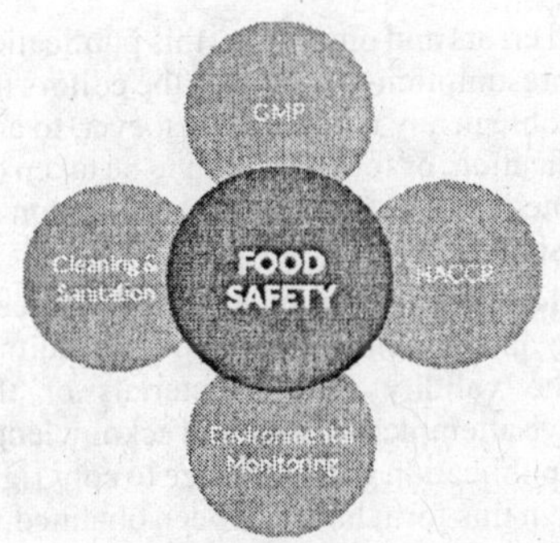

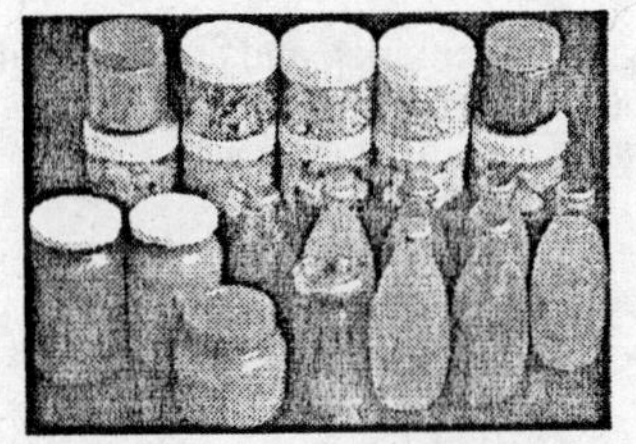

SCIENTIFIC PUBLISHERS

Published by:

SCIENTIFIC PUBLISHERS (INDIA)
5A, New Pali Road, P.O. Box 91
Jodhpur 342 001, India
E-mail: info@scientificpub.com
Website: www.scientificpub.com

ISBN: 978-93-88449-32-8
eISBN: 978-93-88449-33-5

Printed in India

PREFACE

Agricultural production in the country is growing at the rate of 3.5% per annum resulting in marketable surplus in food grains and other agricultural commodities such as horticultural produce, fisheries and animal products etc. This increased agri-horticultural production and animal products would warrant their processing in order to avoid post-harvest losses and at the same time leading to value addition. Thus, here lies the importance of Food Science and Technology discipline in the field of agriculture.

The post harvest losses in the country today are very high (about 30% of total production) in comparison with the post harvest losses (about 5% or less) in most developed countries and some of the developing countries. This is causing tremendous economic loss and making the agricultural activity a commercially unviable proposition to the farmers. The value of post harvest losses has been estimated to be approx. Rs 50,000 crores at national and nearly Rs. 5,000 crores per year at state level. The heavy post harvest losses of fruits and vegetables are attributed to their perishable nature. So, there is an urgent need for post harvest processing of these commodities. Presently, only 2-3% of all agricultural production is processed in the country or state with a low value addition of 35% or so. This is appreciably low in comparison with the reported 50-70% processing and 400-500% value addition in most developed and some of the developing countries around the world. The agro-processing adds value to the produce and helps farmers to get remunerative prices. This can make agriculture a sustainable and a commercially viable activity in the State. Keeping in view the globalization, privatization and liberalization policies in international trade for agriculture based produce and products, the food processing industry will grow and expand rapidly. To operate this sector most efficiently, increased skilled manpower will be required for commercial food processing units, R and D departments, research and educational institutes and quality control departments.

P.M. Kotecha
U.D. Chavan

CONTENTS

1

Food Process Technology

1. HACCP stands for-------------
   (a) Hazard Assessment Critical Control Procedures
   (b) Hazard Assessment Crisis Control Point
   (c) Hazard Analysis Critical Control Point
   (d) Hazard Analysis Critical Control Procedures
2. Preserve foods by using salts and sugars works by--------------
   (a) Raising pH (b) Depleting nutrients
   (c) Lowering osmotic pressure (d) Creating a hypertonic environment
3. Tofu is a processed product of----------------
   (a) Kidney bean (b) Navy bean (b) Broad bean (d) Soybean
4. Which of the molasses contains small amount of complex polysaccharides and invert sugars?
   (a) Cane molasses (b) Beet molasses (c) Invert molasses (d) Blackstrap molasses
5. The pink or reddish liquid that comes from meat on thawing is called----------------
   (a) Drip (b) Leakage (c) Marbeling (d) Oozing
6. Which acid is present in apple?
   (a) Malic acid (b) Citric acid (c) Tartaric acid (d) Oxalic acid
7. CFTRI is situated in--------------
   (a) Kerala (b) Tamil Nadu (c) Karnataka (d) New Delhi
8. A critical control point is an operation by which-------------
   (a) Monitoring is considered unnecessary
   (b) Hazards can be eliminated, minimized or prevented
   (c) Contamination becomes certain
   (d) All risks can be completely eliminated
9. Which one of the following is a step used in setting up a HACCP system?
   (a) Developing control procedures for critical control points
   (b) Taking disciplinary actions if there is a break from procedures

(c) Developing a flowchart for all foods in the operation
(d) Completely eliminating contamination during preparation

10. Which of the following is a function of phytochemicals?
(a) Increase blood lipid values and blood pressure
(b) Diminish the activity of certain enzymes
(c) Increase inflammation
(d) Slow tumor cell growth and instruct cancer cells to die

11. Which of the following is a common example of a functional food?
(a) Beta-carotene in carrots (b) Iodine in salt
(c) Iron in hamburger (d) Zinc in orange juice

12. Which organization is responsible for ensuring that all foods are safe and properly labeled?
(a) FDA (b) WHO (c) NIH (d) CDC

13. Quality as defined by the customer is a/an--------------
(a) Unrealistic definition of quality (b) User-based definition of quality
(c) Manufacturing-based definition of quality (d) Product-based definition of quality

14. Which of the following is not one of the major categories of costs associated with quality?
(a) Prevention costs (b) Appraisal costs (c) Internal failures (d) None of the above

15. According to the manufactures point of view, the definition of quality is ----------
(a) Quality is the degree of excellence at an acceptable price and the control of variability at an acceptable cost
(b) Quality depends on how well the product fits patterns of consumer preferences
(c) Even though quality cannot be defined, you know what it is
(d) Quality is the degree to which a specific product confirms to standards

16. ISO 9000 seeks standardization in terms of -------------
(a) Products (b) Production procedures
(c) Suppliers specifications (d) Procedures to manage quality

17. Which of the following is true about ISO 14000 certification?
(a) It is a prerequisite for ISO 9000 certification
(b) It indicates a higher level of adherence to standards than ISO 9000
(c) It is only sought by companies exporting their goods
(d) It deals with environmental management

18. Total Quality Management emphasizes ---------------
(a) The responsibility of the quality control staff to identify and solve all quality-related problems
(b) A commitment to quality that goes beyond internal company issues to suppliers and customers
(c) A system where strong managers are the only decision makers
(d) A process where mostly statisticians get involved

19. Among the tools of TQM, the tool ordinarily used to aid in understanding the sequence of events through which a product travels is a----------------
(a) Pareto chart (b) Flow chart (c) Check sheet (d) Taguchi map

20. Appertizing is a process called canning and is named after------------
(a) Nicolas Appert (b) Michael Apert (c) John Apert (d) Williams Apert

21. Hedonic rating test is used to measure---------------
(a) Sensitivity (b) Preference (c) Difference (d) Quality

22. Puffed products are dried to a moisture content less than-------------%.
(a) 12 (b) 8 (c) 4 (d) 16

23. During baking, a brown crust develops on the outside of a loaf of bread as the result of---------
(a) Coagulation (b) Denaturation (c) Dextrinisation (d) Emulsification

24. Which of the following food safety issues is a responsibility of state government?
(a) Approving all food safety auditors
(b) Inspecting all food premises annually
(c) Issuing permits for community markets
(d) Developing and updating the Food Standards Code

25. Reverse osmosis is a form of membrane technology that is used to--------------
(a) Cool pasteurized food
(b) Produce some fruit juices
(c) Change the characteristics of some plant foods
(d) Improve the characteristics of some baked products

26. As per BIS, maximum desirable limit of lead in the drinking water is-------------mg/l.
(a) 0.05 (b) 0.09 (c) 0.1 (d) 1.0

27. Zeolite softening process removes------------
(a) Only temporary hardness of water
(b) Only permanent hardness of water
(c) Both temporary and permanent hardness of water
(d) The dissolved gases in permanent hard water

28. Conventional tertiary treatment is-------------
(a) Chemical coagulation and flocculation (b) Filtration
(c) Sedimentation (d) None of the above

29. As per BIS, the maximum desirable limit of total hardness (as $CaCo_3$) in drinking water is-----------------ppm.
(a) 600 (b) 300 (c) 500 (d) 1000

30. Cashew apple is used for making------------
(a) Feni (b) Beer (c) Rum (d) Whiskey

31. The precooling temperature for mango is-------------°C.
(a) 0 (b) 5 (c) 10-12 (d) 18-20

32. The cooked flavour of milk on heating develops due to heat denaturation of ----------------
(a) Lactalbumin (b) Lactoglobulin (c) Casein (d) Lactose

33. ---------- is shaped, fortified and precooked cereal product using continuous press or macaroni press.
(a) TVP (b) Pasta (c) Snacks (d) Breakfast cereal

34. After slaughtering, the stored glycogen of animal is converted into-------------
(a) Glucose (b) Citric acid (c) Carbonic acid (d) Lactic acid

35. The aroma of banana is due to---------------
(a) 2-methyl butyric acid (b) Iso-pentyl acetate
(c) Pentyl valerate (d) Calcium propionate

36. A successful TQM program incorporates all of the following except-----------------
(a) Continuous improvement (b) Employment involvement
(c) Benchmarking (d) Centralized decision making authority

37. Sodium nitrate is-------------
(a) Anticaking agent (b) Antioxidant (c) Curing agent (d) Colorant

38. Which food source below would be the least likely to increase your dietary fiber?
(a) Broccoli (b) Raisin (c) Whole milk (d) White bread

39. Parboiled rice is superior than raw rice because of -------------------
(a) High milling recovery (b) Retention of more protein, vitamins, minerals
(c) More digestibility (d) Increased shelf life

40. Which is not a preservation method in fruit and vegetables?
(a) Pickling (b) Peeling (c) Refrigeration (d) Canning

41. The daily recommendations for dietary fiber are based on----------------
(a) Altitude and lung capacity (b) Activity level and weight
(c) Age and gender (d) Height and weight

42. Which of the following processing methods involves heating foods at high temperatures for short periods in order to reduce the risk of food poisoning?
(a) Blanching (b) Ohmic heating (c) Irradiation (d) Pasteurization

43. Which of the following does not have antimicrobial activity?
(a) Chlorophyll (b) Organic acids (c) Spice extracts (d) Hydrogen peroxide

44. As per FDA, use of hydrogen peroxide has been granted for --------------
(a) Sterilizing equipment (b) Packages in aseptic operations
(c) Both a and b (d) None of the above

45. Which of the following is best to sterilize heat labile solutions?

(a) Autoclave (b) Hot air sterilization
(c) Pasteurization (d) All of these processes kill endospore

46. Ethylene is used for-----------

(a) Retarding ripening of tomatoes (b) Hastening the ripening of fruits
(c) Slowing down ripening of apples (d) Both b and c

47. Organic food production involves--------------

(a) Buying food produced locally
(b) Purchasing food from a farmers market
(c) Growing food using an irrigation system
(d) Growing food without the use of chemical fertilizers and pesticides

48. Thermo resistant bacteria are important in the preservation of foods by-------------

(a) Freezing (b) Canning (c) Chemicals (d) Irradiation

49. Commercial sterilization means-------------

(a) Killing all the organisms
(b) Killing of the organisms including spores
(c) Killing all the organisms with spores except extremely heat resistant spores
(d) None of the above

50. ------------ is the heat treatment used to kill the pathogens.

(a) Pasteurization (b) Tantalization (c) Sterilization (d) Boiling

51. The process of destroying all forms of microorganism is called as------------

(a) Disinfection (b) Sterilization (c) Fermentation (d) None of the above

52. Moist heat sterilization kills the organism by causing--------------

(a) Denaturation of carbohydrates (b) Denaturation of vitamins
(c) Denaturation of vital proteins (d) All of the above

53. Saturated steam is more effective in sterilization than hot air because of ---------------

(a) More penetration (b) Creating vacuum in cells
(c) Making the medium anaerobic (d) None of the above

54. Freeze crack is seen when foods are frozen by-----------------

(a) Air freezing (b) Contact freezing (c) Immersion freezing (d) Cryogenic freezing

55. Which of the following is/are used as refrigerant in cryogenic freezing?

(a) Nitrogen (b) Carbon dioxide
(c) Liquid nitrogen and liquid carbon dioxide (d) All of the above

56. Storage of food material at reduced atmospheric pressure is called as-----------storage.

(a) Controlled atmospheric (b) Modified atmospheric
(c) Hypobaric (d) None of the above

57. Dehydration is a method of preserving food by-------------
    (a) Adding strong concentrations of sugar
    (b) Increasing the salinity level of the food
    (c) Removing moisture by warm air or sunlight
    (d) Heating food in a glass bottle or jar to high temperatures
58. Sweet and salty foods frequently don't require refrigeration to prevent spoilage because they have------------
    (a) Insufficient nutrients (b) Low pH
    (c) High concentration of solutes (d) Naturally occurring antibiotics
59. The quantitative analysis of a food measures its-------------
    (a) Size, shape, flavour, texture and viscosity
    (b) Viscosity, shelf life, nutrient content, weight and volume
    (c) Colour, aroma, mouth-feel, appearance, weight and nutrient content
    (d) Viscosity, flavour, texture, appearance, weight, nutrient content and shelf life
60. CIFT stands for-----------------
    (a) Central Institute of Food Technology (b) Central Institute of Forest Techniques
    (c) Central Institute of Fisheries Technology (d) Central Indian Food Technology
61. ------------- is used for sterilization which works on steam under pressure.
    (a) Hot air oven (b) Boiling water bath (c) Incubator (d) Autoclave
62. The injury caused due to rough surface of containers is known as-------------
    (a) Abrasion injury (b) Chilling injury (c) Chemical injury (d) None of the above
63. Which one of the following is an important guideline for freezer storage?
    (a) Food should be packaged tightly in the freezer
    (b) Freezer temperature should be maintained at 32°F
    (c) Freezer temperature should be maintained at 0°F
    (d) Frozen food inventories do not need to be marked with a date of entry
64. Centrifugation process of removing bacteria from milk is known as--------------
    (a) Milk centrifugation (b) Bactofugation
    (c) Milk fugation (d) Cold centrifugation
65. ----------- is a specific case of food irradiation, where dose of ionizing radiation applied to the food is sufficient to reduce the viable specific pathogen.
    (a) Raduration (b) Radappertization (c) Rad sterilization (d) Radicidation
66. --------- is a process of passing fluid containing suspended particles through a porous medium.
    (a) Sedimentation (b) Centrifugation (c) Extraction (d) Filtration
67. All of the following costs are likely to decrease as a result of better quality except-----------
    (a) Customer dissatisfaction costs (b) Inspection costs
    (c) Scrap costs (d) Maintenance costs

68. What is another name for a functional food?
(a) Supplement (b) Phytochemical (c) Pharmaceutical (d) Nutraceutical

69. The food that has been manipulated to provide additional health benefits is known as-----------
(a) Phytochemical food (b) Functional food
(c) Phamaceutical food (d) Supplement food

70. Which of the following food sources may contain probiotics?
(a) Yogurt (b) Apple slices (c) Whole wheat bread (d) Canned green beans

71. Which of the following conditions would not benefit from the use of probiotics?
(a) Urinary tract infections in women (b) Lactose intolerance
(c) Irritable bowel syndrome (d) Upper respiratory infection

72. Irradiation is also known as--------------
(a) Hot sterilization (b) Pascalization
(c) Cold sterilization (d) None of the above

73. What type of chocolate has the highest flavonoid content?
(a) White
(b) All types of chocolate have the same flavonoid content
(c) Milk
(d) Dark

74. Which of the following may produce free radicals?
(a) Industrial chemicals (d) Clean air
(c) Diet high in fruits and vegetables (d) Moderate exercise

75. What phytochemical is found in a peanut butter and jelly sandwich known for its health-promoting potential?
(a) Resveratrol (b) Lycopene (c) Fiavonoids (d) Lutein

76. Which of the following is a type of phytoestrogen?
(a) Genistein (d) Lutein (c) Indoles (d) Isothiocyanates

77. The compounds found in plants that are believed to have health-promoting effects in humans are called as----------------
(a) Tannins (b) Carotenoids (c) Lycopene (d) Phytochemicals

78. The unstable, highly reactive compounds that damage lipids, protein and DNA in our cells are known as-----------------
(a) Free radicals (b) Antioxidants (c) Phytochemicals (d) Functional foods

79. What phytochemical category would garlic, leeks and onions be classified?
(a) Phytoestrogens (b) Carotenoids
(c) Organosulfur compounds (d) Flavonoids

80. Hardness of water does not----------------
    (a) Have any bad effect in boiler
    (b) Make cooking of foods difficult
    (c) Make it unfit for drinking
    (d) Cause difficulty in the washing of clothes with soaps
81. Permanent hard water may be softened by passing it through----------------
    (a) Sodium silicate (b) Sodium bicarbonate
    (c) Sodium hexametaphosphate (d) Sodium phosphate
82. Temporary hardness of water is caused by the presence of--------------
    (a) Chlorides of calcium and magnesium (b) Sulfates of calcium and magnesium
    (c) Bicarbonates of calcium and magnesium (d) Carbonates of sodium and potassium
83. Permanent hardness of water is due to--------------
    (a) Carbonates in water (b) Bicarbonates in water
    (c) Chlorites and sulphites in water (d) Organic matter in water
84. Phytochemical having a health claim of reducing the risk of age-related eye diseases are-------
    (a) Phenolic acids (b) Flavonoids (c) Phytoestrogens (d) Carotenoids
85. Probiotic cultures----------------
    (a) Encourage the growth of beneficial bacteria in the gut
    (b) Improve sight
    (c) Enhance the body's capacity to absorb calcium
    (d) None of the above
86. Which of the following statements is TRUE regarding probiotics?
    (a) Probiotics are organisms that contribute toward intestinal microbial balance
    (b) Probiotics are non-digestible food products that selectively stimulates the growth of one or a limited number of bacteria in the colon to confer health benefit for the host
    (c) There is good evidence to suggest that probiotics have a beneficial role in preventing post operative recurrence of Crohn's disease
    (d) Milk is an example of a food source containing probiotics
87. Which of the following in an example of a prebiotic?
    (a) Yogurt (b) Inulin (c) Creatinine (d) Fish oil
88. Plant sterols are------------
    (a) A functional ingredient
    (b) Classified as a probiotic substance
    (c) Important in improving the health of the eyes
    (d) Able to stimulate the growth of bacteria in the intestine

89. The process of heating food by use of electric current is known as-------------
(a) Sterilization (b) Pasteurization (c) Ohmic heating (d) All of the above

90. The electromagnetic wave has an energy which is -----------
(a) Inversely related to the wave length (b) Directly related to the wave length
(c) No relation with wave length (d) None of the above

91. Pasteurization is used to destroy ---------------
(a) Antibiotics (b) Bacteria (c) Somatic cells (d) None of the above

92. Taste is ------------- dimensional phenomenon.
(a) Two (b) One (c) Four (d) Three

93. Thixotropy describes a reversible isothermal ----------------- transformation.
(a) Gel – Sol – Gel (b) Sol – Gel – Sol (c) Sol – Gel – Gel (d) Gel – Sol – Sol

94. In duo- trio method of sensory evaluation --------------- samples are used.
(a) One (b) Two (c) Three (d) Four

95. ----------------- is known as queen of spices.
(a) Pepper (b) Cardamom (c) Saffron (d) Ginger

96. Coloured pigment present in turmeric is---------------
(a) Piperin (b) Crocerin (c) Curcumin (d) None of the above

97. ----------------- is part of cinnamon tree which is used as spice.
(a) Leaf (b) Bud (c) Flower (d) Bark

98. Flavour of clove is due to which compound?
(a) Eugenol (b) Anethol (c) Menthol (d) None of the above

99. --------------is actually the dried unopened flower buds.
(a) Ginger (b) Cumin (c) Clove (d) Cardamom

100. ----------------- acid is used in cola type beverages.
(a) Tartaric (b) Citric (c) Mallic (d) Phosphoric

101. Sucrose acetate isobutyrate is --------------- agent.
(a) Sweetening (b) Colouring (c) Flavouring (d) Clouding

102. Spice ----------------- are obtained by solvent extraction with subsequent removal of solvent.
(a) Oils (b) Extracts (c) Oleoresins (d) All of the above

103. In 1767 ---------------- first discovered a method of infusing water with CO_2 to make carbonated water.
(a) Torbern Bergman (b) Dr. Joseph Priestley (c) Michael Owens (d) None of the above

104. Which type of hardness occurs due to the soluble sulphate or chioride of calcium and magnesium?
(a) Permanent (b) Temporary (c) Zero (d) None of the above

105. Saccharin is---------------- times sweeter than sucrose.
(a) 100 (b) 200 (c) 300 (d) 400

106. ------------- is the trade name to the mixture of intensely sweet tasting proteins thaumatin.
(a) Talin (b) Saccharin (c) Aspartin (c) None of the above

107. --------------caramel is used in carbonated beverages.
(a) Bakers (b) Alkali proof (c) Acid proof (d) None of the above

108. ---------------- is used as emulsifier in chocolate manufacturing.
(a) Lecithin (b) Soya protein (c) Starch (d) All of the above

109. Mechanical agitation of coca mass with addition of extra butter is known as --------------
(a) Winnowing (b) Conching (c) Refining (d) Tempering

110. Development of unstable fat crystals on the surface of chocolate is called as --------------
(a) Fat bloom (b) Fat blush (c) Fat spot (c) None of the above

111. Basic concept of HACCP was developed by which company working along with NASA?
(a) Raspberry (c) Blackberry (c) Pillsbury (d) None of the above

112. World Food Day is celebrated on -------------- October.
(a) 13th (b) 14th (c) 15th (d) 16th

113. In lye peeling, fruits and vegetables are dipped in solution of ----------------
(a) HCl (b) Citric acid (c) Water (d) Caustic soda

114. Treatment of fruits and vegetables with boiling water for short period, followed by cooling is called as ----------------
(a) Blanching (b) Pasteurization (c) Exhausting (d) Canning

115. The process of partial sealing of can is called as ----------------
(a) Brining (b) Clinching (c) Grading (d) Exhausting

116. Sulphur resistance lacquered cans are called as----------------can.
(a) R enamel (b) A enamel (c) C enamel (d) M enamel

117. Sparkling clear sweetened fruit juice from which all the pulp and suspended materials are removed is called as----------------
(a) Sherbet (b) Squash (c) Juice concentrate (d) Cordial

118. The phenomenon of spontaneous exudation of fluid from gel is known as -------------
(a) Syneresis (b) Genesis (c) Gelation (d) Synthesis

119. A fruit impregnated with cane sugar and subsequently drained and dried is called as----------
(a) Preserve (b) Candied fruit (c) Fruit powder (d) Syrup

120. A concentrated tomato juice/ pulp without seed, skin and containing not less than 25% tomato solids is called-------------
(a) Tomato ketchup (b) Tomato puree (c) Tomato pulp (d) Tomato paste

121. Vinegar contains ----------- % of acetic acid.
(a) 4 (b) 3 (c) 7 (d) 6

122. Texturized vegetable protein food is a---------------- food.
(a) High fiber high protein (b) Low fiber low protein
(c) High fiber low protein (d) None of the above

123. Natural pigment present in coloured grapes is---------------
(a) Carotenoid (b) Anthocyanin (c) Chlorophyll (d) Zeaxanthin

124. EDTA acts as a--------------
(a) Colouring agent (b) Buffering agent (c) Antioxidant (d) Sequestrant

125. Formation of brown colour in food is due to action of which enzyme?
(a) Phenolase (b) Bromalin (c) Catalase (d) Pectinase

126. Oats are good sources of soluble dietary fibers known as----------------
(a) Cellulose (b) β - glucan (c) Bran (d) Hemicellulose

127. *Idli* mix is a -----------------food.
(a) GM (b) PET (c) Convenience (d) Fast

128. Which additives are essential for the manufacture of whole egg mayonnaise?
(a) Emulsifiers (b) Humectants (c) Antioxidants (d) Bleaching agents

129. Which of the following groups of foods are prepared by fermentation?
(a) Jam, honey, milk (b) Cheese, jam, pickle
(c) Honey, salami, pickle (d) Cheese, yoghurt, salami

130. What is the main reason for packaging bread in plastic?
(a) To maintain air flow (b) To prevent drying out
(c) To improve consumer safety (d) To provide marketing opportunities

131. Deterioration of fats and oils during storage can be minimized by using------------------
(a) Anticaking agent (b) Antioxidant (c) Emulsifiers (d) Preservatives

132. Desirable flavour and aroma of butter is due to-----------------
(a) Diacetyl (b) Lipase (c) Sucrose (d) Lactose

133. Skim milk is whole milk from which ------------- has been removed by centrifugation.
(a) Protein (b) Sugar (c) Water (d) Fat

134. ---------- is a reducing disaccharide in milk which on hydrolysis gives glucose and galactose.
(a) Sucrose (b) Raffinose (c) Lactose (d) Fructose

135. Overrun in ice cream is due to whipping of---------- into semi frozen ice cream mixture which causes swelling of volume.
(a) Water (b) Milk (c) Sugar (d) Air

136. The protein present in egg white is------------------
(a) Gluten (b) Albumen (c) Zein (d) Yolk

137. ------------- is the enzyme present in egg white which is capable of dissolving the bacterial cell wall.
(a) Biotin (b) Avidin (c) Collagen (d) Lysozyme

138. ----------- interaction between muscle protein is responsible for muscle contraction.
(a) Globulin (b) Actomyosin (c) Collagen (d) Elastin

139. ----------- an enzyme from papaya is used for meat tenderization.
(a) Papain (b) Bromalin (c) Lysozyme (d) Invertase

140. Pork is a meat from----------------
(a) Sheep (b) Cow (c) Swine (d) Calf

141. After death of animal, actin and myosin forms stiffness to the muscle which is known as-----
(a) Tenderization (b) Rigormortis (c) Ageing (d) Curing

142. Rice grain contains ----------------% of aleurone layer.
(a) 10-11 (b) 12-14 (c) 20-24 (d) 30-40

143. The outermost covering of rice which is brownish in colour is called as----------------
(a) Aleurone layer (b) Bran (c) Testa (d) Seed coat

144. Cocoa bean is essential ingredient of----------------
(a) Chocolate (b) Milk (c) Flour (d) Both b and c

145. -----------------is product in which liquid phase has a gelling properties.
(a) Fudge (b) Nougat (c) Marshmallow (d) Fondant

146. ------------is an aerated high boiled syrup containing fat.
(a) Marshmallow (b) Nougat (c) Fudge (d) Fondant

147. ----------is essential ingredient for flavour modification.
(a) Salt (b) Sugar (c) Flour (d) Fat

148. Myglobin is a---------------
(a) Pigment (b) RBC's (c) WBC's (d) Colour

149. Veal is obtain from------------------
(a) Pig (b) Goat (c) Buffalo (d) Calve

150. The major inner component of muscle fiber is-----------------
(a) Pigment (b) Heamoglobin (c) Myoglobin (d) None of the above

151. The chief pigment responsible for the pink to red colour of meat muscle is--------------
(a) Hemoglobin (b) Myoglobin (c) Chlorophyll (d) Carotenes

152. Rendering is the process of extraction of -----------------
(a) Protein (b) Carbohydrates (c) Fat (d) Vitamin

153. Castrated male bovine is called as----------------
(a) Calf (b) Steer (c) Vale (d) None of the above

154. Meat is an excellent source of---------------
(a) Mineral (b) Fat (c) Protein (d) None of the above

155. ------------- is flesh or edible part of animal.
(a) Legs (b) Meat (c) Muscle (d) None of the above

156. Meat obtained from domestic pig is called as--------------
(a) Beef (b) Pork (c) Cockerels (d) None of the above

157. --------------is widely used method of food preservation.
(a) Drying (b) Sterilization (c) Irradiation (d) Cooling

158. ------------- is the breakdown of glucose in to pyruvic acid.
(a) Fermentation (b) Glycolysis (c) Putrification (d) None of the above

159. Ethylene is considered as the natural aging and-------------- hormone.
(a) Ripening (b) Inhibiting (c) Destructive (d) None of the above

160. Change in colour of fruits and vegetables occurs specially due to degradation of------------
(a) Chlorophyll (b) Xanthophylls (c) Carotene (d) None of the above

161. Which of the following is not spice?
(a) Ajwan (b) Vanilla (c) Dill (d) All are spices

162. Which of the following is not leafy spice?
(a) Thyme (b) Clove (c) Rosemary (d) Basil

163. Which of the following is sulphur containing spice?
(a) Dill (b) Asafoetida (c) Clove (b) Vanilla

164. Asafoetida is----------------
(a) Root extrudate (b) Leaf (c) Seed (d) Petal

165. Saffron is obtained from----------------
(a) Root (b) Leaf (c) Stigma from flower (d) Seed

166. Bitter component present in saffron is/are---------------
(a) Picrocrocin (b) Safranic acid (c) Safranone (d) All of the above

167. The flavour compound derived from garlic is-----------------
(a) Trimethyl sulphoxide (b) Diallyl sulfide
(c) Tetramethyl sulphonate (d) A mixture of all of the above

168. Major volatile component of saffron aroma is/are--------------
(a) Safranal (b) Safranic acid (c) Safranone (d) All of the above

169. Principle component present in peppermint responsible for typical cooling effect is------------
(a) d-carvone (b) Menthol (c) Linalool (d) None of the above

170. Principle component of pepper is----------------
(a) d-pinene (b) Limonene (c) Eugenol (d) Anethole

171. Eugenol is the principle component of----------------
(a) Clove (b) Coriander (c) Cassia (d) Cardamom

172. Major component responsible for flavour in vanilla is---------------
(a) Vanillin (b) 2-hydroxy vanillin (c) 4-methyl vanillin (d) 2,5-dimethyl vanillin

173. Vanilla flavour is naturally obtained by curing which part of the plant?
(a) Seed (b) Pod (c) Leaves (d) Buds

174. The deep red pungent powder prepared from pods of sweet pepper is------------
(a) Pepper unit (b) Paprika (c) Pepprinol (d) None of the above

175. Chief carotenoids present in chilli is---------------
(a) Xanthophylls (b) Capsanthin (c) Lycopene (d) All of the above

176. Curing is involved in the processing of----------------
(a) Vanilla (b) Green cardamom (c) Saffron (d) All of the above

177. Compounding is typically related with the post harvest processing of------------
(a) Asafoetida (d) Dill (c) Carway (d) All of the above

178. Curcumin, a pigment is found in---------------
(a) Turmeric (b) Ginger (c) Saffron (d) Clove

179. Which of the following spices have highest amount of fiber?
(a) Ginger (b) Rosemary (c) Coriander (b) Clove

180. Uniform black colour of black pepper is generally obtained by--------------
(a) Oven draying (b) Using colour
(c) Enzymatic treatment (d) Black colour is naturally present in pepper

181. Volatile oils can be obtained from spices by using----------------
(a) Microwave (b) Supercritical fluid extraction
(c) Solvent extraction (d) All of the above

182. Corn gluten is rich in the corn protein known as-----------------
(a) Zein (b) Glutenin (c) Glutelin (d) Prolamin

183. Baking powder is of two principle kinds---------------
(a) Less acidic and more acidic (b) Slow and fast acting
(c) Neither a nor b (d) Both a and b

184. Starch granules are present in----------------
(a) Protein matrix (b) Lipid protein complex
(c) Germ (d) None of the above

185. Which wheat variety is most suitable for preparation of biscuits?
(a) Soft wheat (d) Hard wheat (c) Durum wheat (d) All of the above

186. ----------------- is the trade name of ginger oleoresin.
(a) Zingerone (b) Gingerin (c) Gingerol (d) Shogaol

187. Colouring pigment present in turmeric is---------------
(a) Eugenol (b) Curcumin (c) Cineol (d) Terpenol

188. -------------- is a constituent of coffee and tea having bitter taste.
(a) Tannins (b) Caffeine (c) Starch (d) Pentosans

189. --------------- impart yellow to red colour to processed cheese and butter.
(a) Annatto bixin (b) Anthocyanin (c) Lycopene (d) Xanthophyll

190. -----------means forcing material through a die with definite shape to product.
(a) Extrusion (b) Roasting (c) Proofing (d) Puffing

191. -------------- extruder is used for preparation of texturized vegetable protein food.
(a) Collate extruder (b) Twin screw extruder
(c) Low temperature low shear extruder (d) Single screw extruder

192. ----------------- is a bioreactor which works HTST and transforming a variety of raw material into modified intermediate and finished products.
(a) Centrifuge (b) Extruder (c) Electrophoresis (d) Extractor

193. Starch gets ------------ during extrusion cooking.
(a) Caramalize (b) Liquify (c) Gelatinize (d) Gelyfine

194. -----------must be sufficiently strong to tolerate the high pressure during extrusion.
(a) Feeder (b) Barrel (c) Die (d) Screw

195. Chevon is the meat of--------------
(a) Swine (b) Cattle (c) Goat (d) Sheep

196. The process of giving hot water treatment to poultry birds is called as------------
(a) Scalding (b) Stunning (c) Slaughtering (d) Singeing

197. The major component of egg shell is--------------
(a) Phosphorus pentoxide (b) Calcium carbonate
(c) Magnesium carbonate (d) Organic matter

198. Fish liver oil is the principle natural source of------------
(a) Vitamin C (b) Vitamin A (c) Vitamin D (d) Vitamin E

199. What is albumen?
(a) Egg protein (b) Egg pigment (c) Egg white (d) Egg yolk

200. ------------ is a measure of egg protein quality based on height of the egg white.
(a) White index (b) Yolk index (c) White Yolk ratio (d) Haugh Unit

201. The defect sandiness in ice cream is developed because of ------------
(a) Coagulation of casein (b) High lactase content
(c) Low solid contents (d) Slow Freezing

202. Which gas is used as a medium for the preservation of foods?
(a) Nitrogen (b) Oxygen (c) Hydrogen (d) Chlorine

203. The HTST process is advantageous over traditional heat process because-----------
(a) It requires low cost machinery (b) Of complete destruction of microorganism
(c) It is more economical (d) Of less nutrient destruction

204. Phosphatase test is carried out to check the adequacy of-------------
(a) Sterilization (b) Pasteurization (c) Homogenization (d) Clarification

205. *Shrikhand* is the example of ------------ products.
(a) Coagulated (b) Concentrated (c) Fermented (d) Condensed

206. The head quarter of National Dairy Developed Board is situated at ------------
(a) Mumbai (b) Banglore (c) Anand (d) Delhi

207. -------------- is used in dry milling of *tur*.
(a) Red earth (b) Red slurry (c) Oil (d) Water

208. Protein isolates contain about -------------% proteins.
(a) 80-90 (b) 85-90 (c) 80-95 (d) 100

209. ------------- is most widely used solvent in extraction of oil.
(a) Methane (b) Butane (c) Hexane (d) Chloroform

210. Which is a process of removing moisture from foods by sublimation?
(a) Freeze drying (b) Roller drying (c) Spray drying (d) Puff drying

211. ----------- are the most important constituents of jelly.
(a) Sugar (b) Pectin (c) Citric acid (d) All of the above

212. Dried ripe banana is known as--------------
(a) Banana fig (b) Waters (c) Puree (d) Paste

213. Jelmeter test is used for determination of -------------content.
(a) Sugar (b) Pectin (c) Acid (d) Alcohol

214. ------------ is fruit jelly in which the shreds of peel are suspended.
(a) Marmalade (b) Jam (c) Squash (d) Candy

215. TSS of the product can be measured by using -------------
(a) Hand refractometer (b) Jelmeter
(c) Salinometer (d) Viscometer

216. ------------ is a type of pickled cabbage highly popular in Europe and USA.
(a) Canned vegetable (b) Sauerkraut
(c) Acidic pickle (d) Cabbage chutney

217. ------------------ is considered as the natural ripening hormone.
(a) Linolenic acid (b) Ethylene (c) Cytokinins (d) Methionine

218. The loss of green colour of fruits during ripening is due to-------------
(a) Degradation of chlorophyll (b) Lowering TSS
(c) Toughness (d) Increase sugar concentration

219. Blanching is process of inactivation of -----------
(a) Protein (b) Carbohydrates (c) Fats (d) Enzymes

220. Process of heating food after chilling is called as-----------
(a) Dehydration (b) Osmosis (c) Tempering (d) Thawing

221. The process of fermentation produces------------
(a) Alcohol (b) Acid (c) CO_2 (d) All of the above

222. Which of the following method of pasteurization is most recent?
(a) UHT (b) HTST (c) Batch pasteurization (d) In-bottle pasteurization

223. -------------- is natural fondant.
(a) Sugar syrup (b) Glucose syrup (c) Liquid fructose (d) Honey

224. Best flavour in cocoa nibs is developed by process of--------------
(a) Cleaning (b) Drying (c) Washing (d) Alkalization

225. ---------- is common thickening agent in confectionary.
(a) KMS (b) SMS (c) TBHS (d) Pectin

226. -------------- gives bright colour to final confectionary product.
(a) Pectin (b) Lecithin (c) Agar agar (d) Liquid glucose

227. Purple colour in brinjal is due to which pigment.
(a) Anthoxanthin (b) Chrophyll (c) Anthocyanin (d) Carotenoid

228. Respiration Quotient (RQ) value of carbohydrate is------------
(a) >1 (b) 0 (c) <1 (d) 1

229. Which component in chill is responsible for pungency?
(a) Capsaicin (b) Capricin (c) Glycogen (b) Capsailicin

230. Which is climacteric fruit?
(a) Custard apple (b) Grape (c) Ber (d) Amla

231. Generally coloured fruit juices are preserved by which chemical preservative?
(a) KMS (b) Calcium carbide (c) Sodium benzoate (d) Vinegar

232. During cooking, change in colour of meat is due to formation of--------------
(a) Metmyoglobin (b) Myoflobin (c) Oxymyoglobin (b) Actomyosin

233. Fish which migrates from fresh water to marine waters to breed is known as-----------
(a) Anadermous (b) Mesodermus (c) Catadromous (d) None of the above

234. The method used for increasing tenderness of meat is/are-------------
(a) Conditioning or aging (b) Treatment with photolytic enzyme
(c) Mechanical alteration of meat (d) All of the above

235. Meat of hunted animal is known as--------------
(a) Veal (b) Vension (c) Pork (d) None of the above

236. The most common method for pig stunning is--------------
(a) Chemical (b) Electrical (c) Mechanical (d) Both b and c

237. The immediate chilling of fish after harvesting is desirable because--------------
(a) Fish is highly perishable (b) Contain 66-80% water
(c) Rich in unsaturated fat (d) All of the above

238. Which of the following statement is/are correct?
(a) Retrogradation is accelerate by freezing (b) Amylopectin involves in retrogradation
(c) Amylose is having α-1,6 bonding (d) All of the above

239. Product obtained by heating sugar above its melting point in acidic condition is--------
(a) Fructose (b) Carbon dioxide (c) Caramel (d) None of the above

240. The increase in volume, viscosity and translucency of starch granules when they are heated in a liquid is called as ------------
(a) Retrogradation (b) Dextrinization (c) Gelatinization (d) Inversion

241. Starch used in frozen foods should contain------------
(a) High amylase
(b) Less amylose
(c) Amylose content does not make any difference
(d) Less amylopectin

242. Which of the following statement is/ are correct?
(a) Retrogradtion of starch decreases its sweetness
(b) Inversion of the sucrose is isomerization of sucrose
(c) A fluid starch paste is called sol, while a semisolid is known as a gel
(d) All of the above

243. Which of the following is/are responsible for the delay in gelatinization?
(a) Fat (b) Protein (c) Sugar (d) All of the above

244. Which of the following statement is/ are correct?
(a) Sugar competes with starch for available water
(b) Sugar in its higher concentration may even prevent gelatinization process
(c) Sugar increases the required temperature for gelatinization
(d) All of the above

245. When valine is heated with glucose at 180°C, the flavour produced is/are---------
(a) Chocolate (b) Bread like (c) Caramel (d) All of the above

246. Digestion in the mouth and esophagus is a chemical digestion when the--------- glands at the back of the tongue secrete a lingual lipase.
(a) Salivary (b) Adrenaline (c) Ebner's (d) Mucus

247. Drying which involves subjecting the food to a low pressure is called as-----------
(a) Freeze drying (b) Spray drying (c) Puff drying (d) Vacuum drying

248. The process of heating fruit juices for short time at a temperature higher than the pasteurization temperature of the juice is called as-------------
(a) UHT pasteurization (b) HTST pasteurization
(c) Flash pasteurization (d) All of the above

249. The freezing process in which the products are placed in a trays or racks in a long tunnel and cool air is circulated over the product is known as-----------------

(a) Cabinet freezing (b) Tunnel freezing (c) Spiral freezing (d) Cryogenic freezing

250. In HTST process, the milk is pasteurized at--------------

(a) 63°C for 30 min (b) 63°C for 15 sec (c) 72°C for 30 min (d) 72°C for 15 sec

251. The damage to the tissues in fruits and vegetables developed due to refrigeration is called-----

(a) Chilling injury (b) Freezing injury (c) Chemical injury (d) Pathological injury

252. By which of the following, freezing is preceded by partial dehydration---------------

(a) Cryogenic freezing (b) Freeze drying

(c) Vacuum drying (d) Dehydrofreezing

253. Which of the following polymers is/are present in bacterial cell wall?

(a) Techoic acids (b) Polysaccharide (c) Polypeptide (d) All of the above

254. Functions of fat in the body is---------------

(a) Provides a backup energy for the body that can be used when carbohydrates are low

(b) Supplies essential nutrients in the form of fatty acids to the body, which are necessary for proper functioning

(c) Increases one's feeling of fullness after eating

(d) All of the above

255. Which is the essential base for all metabolic processes?

(a) Vitamins (b) Water (c) Carbohydrates (d) Proteins

256. Most of the hydrolysis of triglycerides takes place in---------------

(a) Mouth (b) Stomach (c) Small intestine (b) Large intestine

257. In the ------------, glycogen reserves protect cells from depressed metabolic function.

(a) Bile (b) Liver (c) Small intestine (d) Large intestine

258. Reserve fuel supply and basic fuel supply are the functions of---------------

(a) Fats (b) Proteins (c) Vitamins (d) Carbohydrates

259. Digestion in the small intestine where mechanical digestion rolls food in a spiral motion, exposing new surfaces for absorption by-------------

(a) Peristaltic waves (b) Pendular movements(c) Segmentation ring (d) Longitudinal rotation

260. Which type of carbohydrate is a vital emergency fuel for heart muscle?

(a) Glycogen (b) Triglyceride (c) Glucogenesis (d) Glucose

261. Breaking down of larger substances into smaller units is called as-----------

(a) Catabolism (b) Metabolism (c) Anabolism (b) Botulism

262. Limit all of the following in your diet except---------------

(a) Saturated and trans fats (b) Sugar and salts

(c) Cholesterol (d) Fiber

263. ---------------- food provides substantial nutrients and relatively fewer calories.

(a) Protein (b) Nutrient (c) Nutrient-dense (d) Fatty

264. Foods such as potato chips, tortilla chips, and snack crackers use a fat based replacer called as---- -----------

(a) E-Z Trim (b) Z- E Trim (c) Aspartame (d) Olestra

265. What is the principle behind oven drying for determination of moisture content of foods?

(a) Colour change is measured (b) Loss of weight represents loss of water

(c) Change in refractive index is measured (d) Change in light absorbance is measured

266. Dehydration by soaking fresh material in a heavy sugar or salt solution is known as---------

(a) Osmotic dehydration (b) Atmospheric dehydration

(c) Sub-atmospheric dehydration (d) Vacuum dehydration

267. The storage method in which the gas composition within the storage is not monitored or adjusted is called as------------ storage.

(a) Modified atmospheric (b) Controlled atmospheric

(c) Hypobaric (d) None of the above

268. The storage method in which the altered gas composition inside the storage is monitored and maintained at a preset level is known as------------ storage.

(a) Modified atmospheric (b) Controlled atmospheric

(c) Hypobaric (d) In-situ

269. Storage of food material at reduced atmospheric pressures is called as------------

(a) Controlled atmospheric storage (b) Modified atmospheric storage

(c) Hypobaric storage (d) None of the above

270. A food with a pH of 3.5 is considered to be --------------

(a) Neutral (b) Low acidic (c) High acidic (d) Non-acidic

271. Which of the following chemical is used as a solvent for extraction of fat from foods ample?

(a) Hydrochloric acid (b) Sodium hydroxide

(a) Amylase (d) Ether

272. Which of the following ingredients in chocolate milk comes from seaweed?

(a) Carrageenan (b) Cocoa (c) Sucrose (d) Glucose

273. Which of the following foods is rich in omega-3 fatty acids?

(a) Lard (b) Fish (c) Butter (d) Olive oil

274. Which industrial processing method is most effective for making dried potato flakes?

(a) Drum drying (b) Sun drying (c) Spray drying (d) Osmotic dehydration

275. What happens to the boiling point of water when it is heated at high altitudes?

(a) Increases (b) Decreases

(c) Remains the same (d) Water doesn't boil at high altitudes

276. Which of the following method is a quick test for sugar content during the early stages of the brewing process for beer?

(a) Hydrometry (b) Babcock test (c) Wet-ashing (d) Soxhlet extraction

277. What causes sliced apples to turn brown?

(a) Caramelization (b) Staling (c) Enzyme activity (d) Protein degradation

278. Which of the following does not have antimicrobial activity?

(a) Chlorophyll (b) Organic acids (c) Spice extracts (d) Hydrogen peroxide

279. Which food crop is the richest source of gluten?

(a) Soybeans (b) Corn (c) Rice (d) Wheat

280. What is anethole?

(a) An aroma component (b) A thickening agent

(c) An emulsifier (d) A sweetener

281. Gluten intolerant people can eat--------------

(a) Rye (b) Wheat (c) Bulgur (d) Buckwheat

282. The biological value of foods is based on which main nutrient?

(a) Carbohydrates (b) Fats (c) Protein (d) Dietary fibres

283. What is the name for health-improving bacteria added in dairy products?

(a) Probiotics (b) Prebiotics (c) Synbiotics (d) Symbiotics

284. An enzyme that acts only in an acidic medium is-------------

(a) Pepsin (b) Rennin (c) Trypsin (d) Amylase

285. A non-enzyme protein present in the saliva is----------------

(a) Heparin (b) Mucin (c) Ptyalin (d) None of the above

286. The enzyme that is secreted in an inactive form is----------------

(a) Lipase (b) Trypsin (c) Rennin (d) Ptyalin

287. The digestive juice that is almost neutral in----------

(a) Gastric juice (b) Pancreatic juice (c) Bile juice (d) None of the above

288. ---------------- is the body's primary source of energy.

(a) Fructose (b) Galactose (c) Sucrose (d) Glucose

289. An emulsifying agent produced by the liver and stored in the gall bladder which helps in fat digestion and absorption is-------------

(a) Amino (b) Acid (c) Cholesterol (d) Bile

290. Dietary fiber is an indigestible ---------------- that serves separately as a body regulatory agent.

(a) Fat (b) Protein (c) Carbohydrates (d) Amino acid

291. A nutritionist is a-------------

(a) Registered Dietitian

(b) Certified Diabetes Educator (CDE)

(c) Certified Nutrition Support Dietitian (CNSD)
(d) None of the above

292. In ------------- diet, dairy products are used to complement basic diet of plant foods.
(a) Lacto-ovo vegetarian (b) Lacto-vegetarian
(c) Ovo-vegetarian (d) Vegan

293. Overweight and obesity are major risk factors for -------------- diseases.
(a) Colin cancer (b) Diabetes (c) Lung disease (d) Thyroid cancer

294. Potassium metabisulphite is used as a stable source of------------
(a) Sulphur dioxide (b) Carbon dioxide (c) Sodium benzoate (d) None of the above

295. High carbonation in fruit juices should be avoided as it usually destroys-----------
(a) Colour (b) Flavour (c) Texture (d) All of the above

296. The process of removal of air from cans is known as-------------
(a) Canning (b) Exhausting (c) Sealing (d) Processing

297. Heating of canned foods to kill bacteria is called as--------------
(a) Processing (b) Canning (c) Exhausting (d) All of the above

298. The cans coated with acid resistant lacquer are called as------------cans.
(a) R-enamel (b) C-enamel (c) Sulphur-resistant (d) None of the above

299. Fruit juice from which water has been removed by heating or freezing is known as---------
(a) Concentrate (d) Nectar (c) Cordial (d) Crush

300. Gumminess/stickiness in jam is caused due to-----------
(a) Excess sugar (b) Less sugar (c) Low pectin (d) None of the above

301. The process of removal of moisture under controlled condition of temperature, humidity and air is called as--------------
(a) Drying (b) Dehydration (c) Osmosis (d) Solar drying

302. Minimum TSS content of jam should be------------°B.
(a) 68.5 (b) 69.0 (c) 69.5 (d) 70.0

303. The preservation of food in common salt or in vinegar is known as------------
(a) Fermentation (b) Pickling (c) Pickle (d) Salting

304. The potatoes for chips/wafer making should have-------------
(a) High dry matter and low sugar content
(b) High dry matter and high sugar content
(c) Low dry matter and low sugar content
(d) None of the above

305. The can with mild positive pressure is called as--------------
(a) Flipper (b) Springer (c) Soft swell (d) Hard swell

306. Action of food acids on the can results in a--------------
(a) Hard swell (b) Hydrogen swell (c) Soft swell (d) Nitrogen swell

307. Beer is manufactured from------------

(a) Grape (b) Wheat (c) Barley (d) Jowar

308. The inhibitory action of chemical preservatives is due to their interfering with-----------

(a) The mechanism of cell division (b) Permeability of cell membrane

(c) Activity of enzymes (d) All of the above

309. Destruction of spoilage organisms within the sealed container by means of heat is the principle of-------------

(a) Canning (b) Freezing (c) Drying (d) All of the above

310. Peeling of fruits and vegetables by use of alkali is known as-------------

(a) Hand peeling (b) Mechanical peeling (c) Lye peeling (d) Flame peeling

311. Which of the following is not a fermented beverage?

(a) Champagne (b) Wine (c) Cider (d) Cordial

312. The amount of acid in vinegar is expressed as-------------

(a) Grain strength (b) Brewed vinegar (c) Artificial vinegar (d) Fruit vinegar

313. Radappertization is also known as-------------

(a) Pasteurization (b) Evaporation (c) Sterilization (d) Sprout inhibition

314. Which component is responsible for after cooking blackening of potatoes?

(a) Isochlorogenic acid (b) Chlorogenic acid

(c) Solanin (d) Caffeic acid

315. A group of acidic substances that are produced and secreted as a wound response by specialized plant cells of trees and shrubs are called as-------------

(a) Resins (b) Waxes (c) Emulsions (d) Wood rosin

316. If processed cans are not allowed to cool down sufficiently before storing results in------------

(a) Breathing (b) Bursting (c) Stack burning (d) Flat sour

317. The major quantity of milk in India is utilized for making-------------

(a) Dahi/curd (b) Ghee (c) Butter (d) Ice cream

318. The most variable constituent of milk is------------

(a) Lactose (b) Protein (c) Fat (d) Pigment

319. As per the PFA, the toned milk must have milk fat content of------------

(a) 3.5% (b) 3.0% (c) 4.5% (d) 4.0%

320. The major fraction of milk protein is-------------

(a) α-lactalbumin (b) β-lactoglobin (c) Casein (d) Proteose-peptone

321. The average specific gravity of cow milk is------------

(a) 1.028 to 1.030 (b) 1.030 to 1.032 (c) 1.035 to 1.037 (d) None of the above

322. The adulteration of milk with water is detected by--------------

(a) Acidity (b) Specific gravity (c) Lactometer reading (d) Smell

323. The milk is homogenized to produce-----------------

(a) Soft curd (b) Flavour (c) Britter appearance (d) All of the above

324. According to PFA Rules (1976) the milk cream should contain not less than-----------% fat.

(a) 15 (b) 20 (c) 25 (d) 30

325. The maximum overrun obtainable in butter is--------------%.

(a) 20 (b) 25 (c) 30 (d) 35

326. The requirement of total solids in ice cream is---------------%.

(a) 16 (b) 26 (c) 36 (d) 46

327. A soft, unripened cheese usually made from skim milk is called as-------------

(a) Cheddar (b) Parmesan (c) Gouda (d) Cottage

328. The sweetened condensed milk should contain sugar not less than------------%.

(a) 26 (b) 30 (c) 32 (d) 40

329. The heat stability of milk depends upon-----------------

(a) Calcium + magnesium ions (b) Phosphate + citrate ions

(c) Both a and b (d) None of the above

330. The maximum solubility index of spray dried skim milk powder should be------------

(a) 1.0 ml (b) 2.0 ml (c) 3.0 ml (d) 4.0 ml

331. The base material for preparation of *rossogolla* is-------------

(a) Khoa (b) Paneer (c) Rabri (d) Channa

332. The sandiness in ice cream is caused due to high percentage of---------------

(a) Fat (b) Milk solids not fat (c) Sugar (d) All of the above

333. The object of addition of stabilizer in ice cream is to-------------

(a) Improve whipping quality (b) Give smooth body and texture

(c) Reduce whipping time (d) All of the above

334. After freezing ice cream mix to about 90% overrun, it is hardened at a temperature of---------

(a) –18 to -20°C (b) –15 to -18°C (c) –23 to -29°C (d) –30 to -40°C

335. In cheese making, the casein/fat ratio is standardized to----------

(a) 0.65 to 0.68 (b) 0.68 to 0.70 (c) 0.70 to 0.72 (d) 0.72 to 0.74

336. The object of mass crystallization in cooling of condensed milk is to prevent from------------

(a) Discolouration (b) Age thickening (c) Cooked flavor (d) None of the above

337. The shelf life of paneer at refrigeration temperature is---------------days.

(a) 1-2 (b) 3-4 (c) 5-6 (d) 7-8

338. The coarse texture of channa is caused due to---------------

(a) Excessive acidity in milk (b) Inadequate fat content

(c) Too high temperature of coagulation (d) All of the above

339. The percentage of fat recovery in ghee is highest in which method?

(a) Desi (b) Creamery (c) Direct cream (d) Pre-statification

340. The ghee is usually adulterated by addition of--------------
(a) Vanaspati (b) Refined vegetable oils
(c) Animal fats (d) All of the above

341. The byproducts of Indian dairy industries of paneer, channa and casein is-----------
(a) Butter milk (b) Skim milk (c) Whey (d) None of the above

342. The animals are kept on fast before slaughtering to---------------
(a) Immobilize the animal (b) Maintain glycogen level
(c) Obtain red meat (d) None of the above

343. The most scientific method of slaughtering of animal is--------------
(a) Kosher (b) Zatka (c) Halal (d) None of the above

344. The ultimate pH of the fresh meat is--------------
(a) 4.6 (b) 5.6 (c) 6.6 (d) 7.6

345. The meat is cured by using---------------
(a) Nitrates (b) Nitrites (c) NaCl (d) All of the above

346. -------------- is the absence of infectious microorganisms in living tissue.
(a) Sterilization (b) Asepsis (c) Disinfection (d) Microbiostasis

347. The egg quality is assessed by--------------
(a) Candling (b) Albumen index (c) Haugh unit (d) All of the above

348. The mostly followed preservation method of egg at farm levels is---------------
(a) Lime water (b) Coating of egg with oil
(c) Thermostabilization (d) Freezing

349. The isoelectric point of milk casein is------------
(a) 3.5 (b) 4.0 (c) 4.7 (d) 5.0

350. In myoglobin and hemoglobin, the porphyrins are held by metal------------
(a) Mg (b) Fe (c) Cu (d) Zn

351. The surface reddening in canned meat is due to-----------
(a) Nitrates (b) Nitrites (c) Carbon monoxide (d) All of the above

352. The colostrum contains approximately--------------% protein.
(a) 17.5 (b) 10.5 (c) 5.5 (d) 15.5

353. The most abundant protein in egg white is-----------
(a) Ovomucoid (b) Conalbumin (c) Ovalbumin (d) Lysozymes

354. The spoiled fishes are used for preparation of-----------
(a) Strained baby food (b) Fish protein concentrate
(c) Fish meal (d) None of the above

355. The cooked flavour of milk on heating is due to denaturation of----------
(a) Lactalbumin (b) Lactoglobulin (c) Casein (d) Lactose

356. The milk serum or whey is that portion of milk which is devoid of-------------
(a) Casein (b) Fat (c) Both a and b (d) Lactose

357. The flatulence producing oligosaccharides in legumes are effectively reduced by--------------
(a) Cooking (b) Frying (c) Sprouting (d) Roasting

358. In textural properties, gumminess is term applied for----------------
(a) Semi solid products (b) Solid products
(c) Liquid products (d) All of the above

359. Pomegranate export facility center is located at--------------
(a) Mangalvedha (b) Sangamner (c) Amaravati (d) Baramati

360. The range of visible wavelength is-------------
(a) 380-770 mμ (b) 770-840 mμ (c) 280-380 mμ (d) None of the above

361. When light is partially absorbed the object appears------------
(a) Black (b) White (c) Gray (d) Red

362. Purity in terms of sensory terms means----------------
(a) Chroma (b) Intensity (c) Strength (d) All of the above

363. Which one of the following is not mouth feel property?
(a) Fibrousness (b) Mealiness (c) Stickiness (d) Juiciness

364. E-nose is designed for detection of---------------
(a) Volatile organic compounds (b) Non-volatile organic compounds
(c) Both a and b (d) None of the above

365. Which one of the following is not animal based clarifying agent-------------
(a) Egg white (b) Gelatin (c) Bentonite (d) Icing glass

366. Cashew apple is used for making---------------
(a) Feni (b) Beer (c) Rum (d) Whiskey

367. Rice husk is rich source of---------------
(a) Iron (b) Mica (c) Silica (d) Phosphorus

368. The byproduct of fish processing industry used as fining agent is------------
(a) Scales (b) Gelatin (c) Fish meal (d) Fish oil

369. Cane molasses a byproduct of sugar industry contains-----------% sugar.
(a) 50-55 (b) 40-50 (c) 35-40 (d) 25-30

370. The crown of pineapple is used for extraction of---------------
(a) Bromelin (b) Catalase (c) Gluco amylase (d) Papain

371. The UV spectrum has wavelengths between---------------
(a) 50-100 nm (b) 100-400 nm (c) 400-700 nm (d) None of the above

372. Cholesterol can act as a precursor for each of the following compounds except-------------
(a) Chenodeoxycholic acid (b) Testesterone
(c) Glycocholic acid (d) Cholecystokinin

373. Which of the following type of wheat is used in bread making?
(a) Triticum aestivum (b) Triticum durum
(c) Triticum compactum (d) Triticum vulgare

374. The preparation of protein isolates from legume/oilseed meals involve---------------
(a) Acidic extraction followed by isoelectric precipitation
(b) Alkaline extraction followed by isoelectric precipitation
(c) Neutral extraction followed by isoelectric precipitation
(d) Alkaline extraction followed by salt precipitation

375. Damaged starch in bread flour should be-------------%.
(a) 6 (b) 7 (c) 9 (d) 11

376. Fruits and vegetables are excellent dietary source of---------------
(a) Carbohydrates (b) Minerals and vitamins
(c) Protein (d) Fat

377. The end product of ethanol oxidative fermentation is--------------
(a) Lactic acid (b) Pyruvic acid (c) Ethanol (d) Acetic acid

378. Which of the following is a non-climacteric fruit?
(a) Mango (b) Orange (c) Banana (d) Sapota

379. Lactose intolerance observed in some infants is due to--------------
(a) Genetic error (b) Poor health of mother
(c) Poor health of infant (d) Defective milk of mother

380. Excellent dietary source of vitamin A is--------------
(a) Mango (b) Carrot (c) Leafy vegetables (d) All of the above

381. Rigor mortis which develops soon after slaughter of animal is due to---------------
(a) Loss of ATP (b) Decrease in pH
(c) Failure of blood circulation (d) All of the above

382. Removal of tannins from fruit juices can be achieved by treatment with---------
(a) Polyvinyl pyrrolidone (b) Bentonite
(c) Pectinol (d) Papain

383. The pH of fruit jelly is always kept at 2.8 to 3.5 to facilitate------------
(c) Improvement in flavour (b) Rapid cooking
(c) Formation of stable gel (d) Dissolution of more surcrose

384. The widely used preservative to control enzymatic browning in fruits and vegetables during processing and drying is ---------------
(a) Potassium permanganate (b) Potassium metabisulfite
(c) Sodium chloride (d) Organic acids

385. Which of the following acids is predominant in tamarind?
(a) Citric (b) Malic (c) Tartaric (d) Oxalic

386. Which is the major flavour component of vinegar?
(a) Acetic acid (b) Esters (c) Citric acid (d) Ethers

387. Synthetic sweetners are commonly used as sweetening agents in food because------------
(a) They are easily available
(b) The sweetness response is faster
(c) They have long shelf life
(d) They have low calorific value per unit of sweetness

388. Sodium nitrite is used in curing of meat to improve---------------
(a) Colour (b) Flavour (c) Texture (d) Taste

389. Which of the following is used as leavening acid in preparation of baking powder?
(a) Potassium acid tartarate (b) Citric acid monocalcium phosphate
(c) Citric acid monohydrate (d) Stearyol-2-lactylate

390. The oxidizing agent extensively used to improve dough properties of wheat maida is--------
(a) Potassium bromate (b) Potassium dichromate
(c) Potassium permagnate (d) Potassium iodate

391. The most abundant unsaturated fatty acid in cow milk fat is-------------
(a) Oleic (b) Linoleic (c) Searic (d) Pamitic

392. Which pattern of respiratory activity is exhibited by vegetables?
(a) Climacteric (b) Preclimacteric (c) Non-climacteric (d) None of the above

393. Which is the major fraction of maize protein?
(a) Glutelin (b) Gliadin (c) Gluten (d) None of the above

394. Which is an effective antimutagen phenolic constituent present in strawberries?
(a) Tannin (b) Ellagic acid (c) Citric acid (b) Anthocyanin

395. The most predominant fatty acid in sunflower oil is--------------
(a) Oleic (b) Linolenic (c) Linoleic (d) Arachiodonic

396. The colour of meat turns brown upon cooking due to-----------
(a) Oxidation of iron in myoglobin (b) Reduction of iron in myoglobin
(c) Degradation of myoglobin (d) Interaction of myoglobin with haemoglobin

397. Ripping a part of curd by rapid gas production is called as--------------
(a) Sweet curdling (b) Putrefation
(c) Stormy fermentation (b) Ropiness

398. Which of the following fruit is climacteric?
(a) Orange (b) Lemon (c) Mango (d) Grape

399. What temperature should be maintained for transport of banana by ship?
(a) 4.4°C (b) 14.4°C (c) 24.4°C (d) 34.4°C

400. Which of the following vegetables is a good source of oleoresin?
(b) Beet (b) Tomato (c) Paprika chilli (d) Carrot

401. Which of the following sugars has highest sweetness?
(a) Sucrose (b) Fructose (c) Glucose (d) Xylose

402. Gluten in wheat flour is made of gliadins and-----------
(a) Albumins (b) Globulins (c) Prolamins (d) Glutenins

403. Rice bran undergoes spoilage very quickly because of -------------
(a) Active lipase in bran (b) Quick microbial attack
(c) Quick moisture uptake (d) Oxidation of bran oil

404. Which of the following treatment are used in purification of water?
(a) Physical treatment (b) Chemical treatment
(c) Biological treatment (d) All of the above

405. Deodorization of cream by steam distillation under reduced pressure is called as-----------
(a) Pasteurization (b) Vacreation (c) Sterilization (d) Thermosterilization

406. How many parts by weight of 40% cream and 3% milk must be mixed to make milk testing 5% fat?
(a) 1.0 (b) 2.0 (c) 3.0 (d) 4.0

407. The milk serum or whey is that portion of milk which is devoid of-------------
(a) Casein (b) Fat (c) Both a and b (d) Lactose

408. Gariatic foods are designed for-------------
(a) Pregnant women (b) Infants (c) Diabatic people (d) Older people

409. The residual level of SO_2 in raisin permitted in India is--------------ppm.
(a) 250 (b) 500 (c) 750 (d) 1000

410. The irradiation center for mango is located at --------------
(a) MPKV, Rahuri (b) Nasik (c) Lasalgaon (d) Talegaon

411. The purple colour of jamun fruits is due to ---------------
(a) Anthocyanins (b) Lycopene (c) Cryptoxanthin (d) Querecitin

412. Tomato puree should have minimum TSS of--------------°B.
(a) 5 (b) 7 (c) 9 (d) 12

413. The egg white constitutes to about--------------%.
(a) 59 (b) 29 (c) 39 (d) 19

414. A semi permeable membrane allows the flow of---------------
(a) Solvent molecules (b) Solute molecules
(c) Both solute and solvent (d) None of the above

415. Which one of the following is not a chemical preservative?
(a) Acetic acid (b) SO_2 (c) Propionic acid (d) Dextran

416. The process of removal of odoriferous and flavoured substances in oil is known as-----------
(a) Degumming (b) Bleaching (c) Deodourization (d) Refining

417. The process of removal of natural colouring matter in oil is known as------------
(a) Degumming (b) Bleaching (c) Deodourization (d) Refining

418. The process of exposing the dough to 31-32^{0}C with 80% RH in closed chamber to increase the loaf volume of dough is known as-------------
(a) Proofing (b) Crusting (c) Knock back (d) Baking

419. The process of saponifying free fatty acids in oil with aqueous NaOH is known as------------
(a) Neutralization (b) Saponification (c) Bleaching (d) Deodourization

420. The active principle compound in garlic which is having antibiotic properties is-------------
(a) Tannin (b) Allicin (c) Caffein (d) Gingerin

421. Ragi is rich source of---------------
(a) Calcium (b) Iron (c) Fat (d) Both a and b

422. The principle protein of rice is-----------------
(a) Gluten (b) Zein (c) Oryzenin (d) None of the above

423. Triticale is a hybrid cereal from a cross between--------------
(a) Wheat and rye (b) Wheat and ragi (c) Rye and barley (b) Ragi and barley

424. Among the cereals -------- is the unique as it contains high level of soluble dietery fibres.
(a) Wheat (b) Ragi (c) Barley (b) Bajra

425. Oils from cereal grains are rich in which vitamin?
(a) A (b) B (c) E (d) D

426. Rice is used in the preparation of---------------
(a) Gin (b) Beer (c) Sake (b) Brandy

427. Wet milling tends to produce --------------- products.
(a) Coarser (b) Medium size (c) Finer (d) None of above

428. The paddy grain is mainly composed of --------------- starch granules.
(a) Hexagonal (b) Polygonal (c) Tringular (d) Circular

429. The differential speed ratio of break rolls of wheat mill is about--------------
(a) 1:1 (b) 2:1 (c) 2.5:1 (d) 1:2

430. Coarsely ground white endosperm chemically same to white flour is known as --------------
(a) Semolina (b) Flour (c) Mill feed (d) Shorts

431. Mixture of bran, germ and shorts obtained during milling is called as ------------
(a) Flour (b) Mill feed (c) Semolina (d) Mill shorts

432. In which of the following process moisture content in the grain is adjusted?
(a) Conditioning (b) Parboiling (c) Milling (d) Parching

433. Mixture of bran and germ produced from milled wheat is known as---------------
(a) Mill feed (b) Wheat shorts (c) Flour (d) Semolina

434. Rheology is the study of -------------- properties of the dough.
(a) Elastic (b) Textural (c) Viscoelastic (d) Both a and b

435. Loss of moisture by evaporation during milling is known as-------------
(a) Milling drying (b) Milling loss (c) Moisture loss (d) All of the above

436. ----------- are important in determining gluten characteristics.
(a) Glutamine and gliadine (d) Gliadine and glutenine
(c) Aspargine and gliadine (d) Glutamine and glutenine

437. Which type of wheat is used for preparation of semolina?
(a) Aestivum (b) Dicoccum (c) Durum (d) Compactum

438. For manufacture of biscuits, which type of wheat is used?
(a) Soft wheat (b) Hard wheat (c) Durum (d) All of the above

439. Flour and egg used as baking ingredients acts as---------------
(a) Tenderizers (b) Toughners (c) Structure builder (d) Both b and c

440. Whole wheat flour is also known as--------------
(a) Perfect flour (b) Straight flour (c) Graham flour (d) None of the other

441. ------------- generally used as the anti staling agent in bread.
(a) GMS (b) GSM (c) MSG (d) SGM

442. Gliadine fraction of protein in gluten is responsible for ---------------
(a) Water holding capacity (b) Extensibility
(a) Elasticity (d) All of the above

443. Food products under dough system includes-------------
(a) Bread (b) Pasta products (c) Cookies (d) All of the above

444. ------------- are the food ingredient that improves the conditions in the large intestine beneficially for host.
(a) Probiotics (b) Prebiotics (c) Synbiotics (d) Antibiotic

445. Food or naturally occurring food supplement which is having disease preventive and health promoting properties are called as-------------
(a) GM food (b) Neutraceuticals (c) Antioxidant (d) None of the above

446. Weaning food is specially formulated for infants aged between -------------months.
(a) 1-2 (b) 3-9 (c) 16-18 (d) 21-24

447. Prebiotics are mainly --------------- carbohydrates.
(a) Digestible (b) Non digestible (c) Soluble (d) Crude

448. Body needs more fluid at-------------
(a) Sea level (b) High attitude (c) Low attitude (d) None of above

449. ------------- is phytochemical found in tomatoes.
(a) Lycopene (b) Quercetin (c) Flavonoids (d) Catechin

450 ---------------- is the study of foods and food constituents on gene expression.
(a) Pretromics (b) Nutrigenomics (c) Nutrition (d) None of the above

451. NHP stands for----------------
(a) Natural Healing Product (b) Natural Health Product
(c) Nutritional Health Programme (d) Non Hygienic Practices

452. Omega-3 fatty acid, the neutraceutical from -------------- source.
(a) Plant (b) Animal (c) Microbial (d) None of the above

453. Hydrolysis of triglycerides with alkali is called as-------------
(a) Titration (b) Hydrogenation (c) Saponification (d) Rancidity

454. Iodine value measures--------------
(a) Degree of unsaturation (b) Degree of saturation
(a) Amount of carbon present (d) No. of iodine present

455. For hydrogenation of oils, the catalyst required is--------------
(a) Iron (b) Aluminium (c) Nickel (d) Magnesium

456. Agar is a --------------
(a) Polypeptide (b) Polysaccharide (c) Polyphenol (d) Polyflavone

457. Anthocynins are stable at------------
(a) Neutral pH (b) Acidic pH (c) Basic pH (d) Independent of pH

458. After the removal of Mg ion from chlorophyll, the compound formed is --------------
(a) Pheophytin (b) Chlorophillide (b) Pheophorbide (d) Pyrophorbide

459. Which of following is / are used as the antioxidants in fats and oil processing?
(a) Calcium propionate (b) Butylated hydroxyl hydrazine
(b) Butylated hydroxyl anisole (d) All of the above

460. The product of enzymatic browning is-------------
(a) Melanin (b) Melanoidins (c) Caramel (d) All of the above

461. Which one of the adjunct contributes for foam stability?
(a) Rice (b) Syrups (c) Wheat (d) Corn

462. Each gram of alcohol contains -------------- calories.
(a) 4 (b) 5 (c) 6 (d) 7

463. The degradation /oxidation of alcohol by liver is--------------units/hr.
(a) 1 (b) 2 (c) 3 (d) 4

464. During wort boiling what happens to proteins?
(a) Coagulation (b) Hydrolysis (c) Solubilization (d) All of the above

465. Milk fat is present in ------------ forms.
(a) Colloidal (b) Emulsion (c) True solution (d) None of theses

466. Which amino acid is used as flavour enhancer in food industry?
(a) Citric acid (b) Glutamic acid (c) Aspartic acid (d) All of the above

467. Ghee contains ---------------- % minimum fat.

(a) 60 (b) 99.5 (c) 50 (d) 40

468. The nature of freshly drawn milk is-------------

(a) Acidic (b) Basic (c) Amphoteric (d) Neutral

469. The natural acidity of milk ranges between------------%.

(a) 0.13-0.15 (b) 0.05-0.10 (c) 0.25-0.50 (d) 0.21-0.25

470. A cryoscopy is an important tool that detects -------------- in milk.

(a) Butterfat (b) Antibiotics (c) Pesticides (d) Added water

471. The yellow colour of milk is due to--------------

(a) Anthocyanin (b) Carotene (c) Betalin (d) Riboflavin

472. Egg provides -------------kcals/100g.

(a) 90 (b) 47 (c) 173 (d) 200

473. Lyophilization is also known as------------

(a) Freeze drying (b) Drying under vacuum

(c) Drying only (d) None of the above

474. Collagen on heating in presence of moisture dissolves and yields -------------

(a) Alginate (b) Gelatin (c) Pectin (d) Casein

475. Cooking of fish requires a much shorter time than meat and poultry because ------------

(a) Fish has little connective tissue

(c) Fish has more connective tissues

(d) Fish contains less amount of fat than meat and poultry

(c) Both a and c

476. Poor quality egg floats in the water due to-------------

(a) Microbial spoilage (b) Increase in air cell

(c) Decrease in air cell (d) Air cell has no relation to quality of egg

477. The biological value of fish protein is ---------------

(a) 74 (b) 66 (c) 96 (d) 80

478. The brown discolouration caused by mold growth in salted fish is termed as-------------

(a) Dun (b) Drip (c) Curing (d) Dill

479. The anticoagulant Heparin is obtained from --------------- of animals.

(a) Liver (b) Lungs (c) Heart (d) Kidney

480. National Research Center on meat is located at---------------

(a) Hyderabad (b) Chennai (c) Delhi (d) Mumbai

481. Glazing of fish is done to protect fish from ---------------

(a) Microbial spoilage (b) Freezer burn

(c) Oxidation and freezer burn (d) Chemical spoilage

482. The enzymatic activity of egg yolk is associated with-------------
(a) Phosvitin (b) Livetin (c) Lipovitellin (d) None of the above

483. The brine solution is prepared by using-----------
(a) Sugar (b) KMS/NaMS (c) Salts (d) None of the above

484. ------------removes permanent hardness.
(a) Soda ash (b) Zeolite (c) Resins (d) Lime

485. ------------- gives rigidity to cell wall.
(a) Peptidoglycan (b) Lipid (c) Cell membrane (d) Plasma membrane

486. Filter sterilization is a---------------- method of sterilization.
(a) Heat (b) Stream (b) Physical (d) None of the above

487. The formation of hard permeable skin on the surface of fruits and vegetables during drying is known as------------
(a) Shrinkage (b) Fluidization (c) Case hardening (d) None of the above

488. --------------- is the concentration of solution by boiling of solvent.
(a) Extraction (b) Expression (c) Evaporation (d) Dehydration

489. ----------------- is commonly used for drying vegetable seeds.
(a) Tray drying (b) Fluidized bed drying
(c) Osmotic drying (d) Radiation drying

490. Which of the following method of pasteurization is most recent?
(a) UHT (b) HTST
(c) Batch pasteurization (d) In-bottle pasteurization

491. -------------- is a process of filtration, in which solid particles are separated from a fluid by passing through a porous membrane.
(a) Ultrafiltration (b) Centrifugal filtration
(c) Reverse osmosis (d) All of the above

492. --------- is the heating of electrically non conducting materials by rapidly varying electromagnetic field.
(a) Microwave heating (b) Dielectric heating
(c) Steam heating (d) None of the above

493. - - - - - are regarded as those electromagnetic radiations having frequencies in the range of 3-3x105 MHZ.
(a) Infrared (b) X ray (c) Microwave (d) All of the above

494. The foods of irregular shape and individually quick frozen products are produced by --------- method of freezing.
(a) Freezing by contact with a cooled solid
(c) Freezing by contact with a cooled liquid

(c) Freezing by contact with a cooled gas
(d) All of the above

495. The fish, meat, poultry and eggs are frozen by the ----------- method.
(a) Sharp freezing
(b) Quick or rapid freezing
(c) Blast freezing
(d) None of the above

496. ---------- of foods results in savings in weight ,volume and shelf life of fresh material.
(a) Rehydration
(b) Dehydration
(c) Extraction
(d) Evaporation

497. Maximum carbohydrates are obtained from-------------
(a) Whole grain food
(b) Fatty fish
(c) Plant oil
(d) Nuts

498. According to food guide pyramid fats oils and sweets should be used---------
(a) 2-3 servings.
(b) Frequently
(c) Sparingly
(d) 2-4 servings

499. Tryptophan is--------------
(a) Essential fatty acid
(b) Non-essential acid
(c) Essential amino acid
(d) Non-essential amino acid

ANSWER KEY

1. c	2. d	3. d	4. d	5. a	6. a	7. a	8. b	9. a	10. d
11. b	12. a	13. b	14. d	15. d	16. d	17. d	18. b	19. b	20. a
21. b	22. b	23. c	24. a	25. b	26. a	27. c	28. a	29. b	30. a
31. c	32. a	33. b	34. d	35. a	36. d	37. c	38. c	39. b	40. b
41. c	42. d	43. a	44. c	45. d	46. b	47. d	48. b	49. c	50. a
51. b	52. c	53. a	54. d	55. c	56. c	57. c	58. c	59. a	60. c
61. d	62. a	63. c	64. b	65. d	66. d	67. d	68. d	69. b	70. a
71. d	72. c	73. d	74. a	75. a	76. a	77. d	78. a	79. c	80. c
81. c	82. c	83. c	84. d	85. a	86. a	87. a	88. a	89. c	90. a
91. b	92. c	93. a	94. c	95. b	96. c	97. d	98. a	99. c	100. d
101. d	102. c	103. b	104. a	105. c	106. a	107. c	108. a	109. b	110. a
111. c	112. d	113. d	114. a	115. b	116. c	117. d	118. a	119. b	120. d
121. a	122. a	123. b	124. d	125. a	126. b	127. c	128. a	129. d	130. b
131. b	132. a	133. d	134. c	135. d	136. b	137. d	138. b	139. a	140. c
141. b	142. b	143. b	144. a	145. c	146. b	147. a	148. a	149. d	150. c
151. b	152. c	153. b	154. c	155. b	156. b	157. a	158. b	159. a	160. a
161. d	162. b	163. b	164. a	165. c	166. a	167. b	168. a	169. b	170. a
171. a	172. a	173. b	174. b	175. b	176. d	177. a	178. a	179. a	180. c
181. d	182. a	183. b	184. a	185. a	186. b	187. b	188. a	189. a	190. a

191. d	192. b	193. c	194. b	195. c	196. a	197. b	198. b	199. c	200. d
201. b	202. a	203. d	204. b	205. c	206. c	207. a	208. b	209. c	210. a
211. d	212. a	213. b	214. a	215. a	216. b	217. b	218. a	219. d	220. d
221. d	222. a	223. d	224. d	225. d	226. d	227. c	228. d	229. a	230. a
231. c	232. a	233. c	234. d	235. b	236. a	237. d	238. a	239. c	240. c
241. b	242. c	243. d	244. d	245. a	246. c	247. d	248. c	249. b	250. d
251. a	252. d	253. d	254. d	255. b	256. c	257. b	258. d	259. d	260. a
261. a	262. d	263. c	264. d	265. b	266. a	267. a	268. b	269. c	270. c
271. d	272. a	273. b	274. a	275. b	276. a	277. c	278. d	279. d	280. a
281. d	282. c	283. a	284. a	285. b	286. b	287. b	288. d	289. d	290. c
291. d	292. b	293. b	294. a	295. b	296. b	297. a	298. a	299. a	300. a
301. b	302. a	303. b	304. a	305. a	306. b	307. c	308. d	309. a	310. c
311. d	312. a	313. c	314. b	315. a	316. c	317. b	318. c	319. b	320. c
321. a	322. c	323. d	324. c	325. b	326. c	327. d	328. d	329. c	330. b
331. d	332. b	333. d	334. c	335. b	336. b	337. c	338. d	339. c	340. d
341. c	342. b	343. c	344. b	345. d	346. b	347. d	348. a	349. c	350. b
351. d	352. a	353. c	354. c	355. a	356. c	357. c	358. a	359. d	360. a
361. c	362. d	363. d	364. a	365. c	366. a	367. c	368. b	369. a	370. a
371. b	372. d	373. a	374. b	375. c	376. b	377. d	378. b	379. a	380. d
381. d	382. a	383. c	384. b	385. c	386. b	387. d	388. a	389. c	390. a
391. a	392. c	393. a	394. b	395. c	396. a	397. c	398. c	399. b	400. c
401. b	402. d	403. a	404. d	405. b	406. b	407. c	408. d	409. c	410. c
411. a	412. c	413. a	414. a	415. d	416. c	417. b	418. a	419. a	420. b
421. d	422. c	423. a	424. c	425. c	426. c	427. c	428. b	429. c	430. a
431. b	432. a	433. b	434. c	435. b	436. b	437. d	438. a	439. d	440. c
441. c	442. b	443. d	444. b	445. b	446. b	447. b	448. b	449. a	450. b
451. b	452. b	453. c	454. b	455. c	456. b	457. a	458. a	459. c	460. a
461. c	462. d	463. a	464. a	465. b	466. b	467. b	468. c	469. a	470. d
471. b	472. c	473. a	474. b	475. a	476. b	477. d	478. a	479. b	480. a
481. c	482. b	483. c	484. d	485. a	486. c	487. c	488. c	489. b	490. a
491. a	492. b	493. c	494. c	495. a	496. b	497. a	498. c	499. c	

Food Engineering

1. The heat transfer is constant when ------------
   (a) Temperature remains constant with time (b) Temperature increases with time
   (c) Temperature decreases with time (d) None of the above
2. Heat transfer by conduction follows ------------
   (a) Stepson's law (b) Newton's law
   (c) First law of thermodynamics (d) Fourier's law
3. The term is $\frac{\Delta x}{KA}$ called --------------
   (a) Thermal resistance of wall (b) Thermal conductance of the wall
   (c) Heat flux density (d) Thermal loading of the wall
4. The term $\frac{hl}{K}$ is called-------------
   (a) Reynolds number (b) Prandtl number
   (c) Froude number (d) Nusselt number
5. Thermal conductivity of water-----------
   (a) First increases with temperature then decreases with temperature
   (b) Increases steadily with temperature
   (c) Decreases with temperature
   (d) Does not depend on temperature
6. The term $\frac{\sigma lv}{\mu}$ is called-------------
   (a) Froude number (b) Reynolds's number (c) Prandtl number (d) Mach number
7. Rotameter is a device to measure -------------
   (a) Pressure of fluid (b) Velocity of fluid
   (c) Cross sectional area of fluid (d) Flow rate of fluid
8. Pump is a device to convert--------------
   (a) Electrical energy in to mechanical energy (b) Mechanical energy into hydraulic energy
   (c) Hydraulic energy into mechanical energy (d) Electrical energy into hydraulic energy

9. Centrifugal pump are used for --------------
   (a) Low pressure head (b) High pressure head
   (c) Constant pressure head (d) None of the above
10. Gauge pressure is the pressure------------
   (a) Above the absolute pressure (b) Below the atmospheric pressure
   (c) Equal to atmospheric pressure (d) None of the above
11. Plant location decision is------------
   (a) Short term decision (b) Long term decision
   (c) Constant term decision (d) Depend upon the product
12. What is the important objectives of plant layout?
   (a) Maximum material handling (b) Minimum material handling
   (c) Both a and b (d) None of the above
13. Excessive corrosion -------------- the shelf-life of construction material.
   (a) Increases (b) Decreases (c) Maintains (d) Does not effect
14. In product layout --------------- type of products is in operating area.
   (a) One (b) Two (c) Heavy (d) Weight less
15. Product layout is suitable for ---------------- process industries.
   (a) Batch (b) Continuous (c) Group (d) All the above
16. When two or more materials are mixed to get a product, the process is known as------------- process.
   (a) Analytical (b) Synthetic (c) Conditioning (d) Extractive
17. If the products are too heavy -------------- types of layout are used.
   (a) Line (b) Product (c) Fixed position (d) Group
18. Which type of technique is used for production, planning and control?
   (a) Travel chart (c) Cross chart
   (c) Activity relationship chart (d) Flow process chart
19. In process arrangement ------------ technique are used.
   (a) Spiral method (b) Assignment (c) Travel chart (d) Straight line
20. What is entropy?
   (a) The amount of energy in a system (b) The amount of work done
   (c) The disorder in a system (d) The efficiency of a system
21. Apples are generally wrapped in waxed paper to-----------
   (a) Prevent sunlight for changing its colour
   (b) Prevent aerobic respiration by checking the entry of O_2
   (c) Prevent ethylene formation due to injury
   (d) Make the apples look attractive

22. A system which permits the passage of both matter and energy across it, is called as--------- system.
(a) Closed (b) Open (a) Adiabatic (d) Isolated

23. An unsaturated solution------------
(a) Hasn't dissolved as much solute as is theoretically possible
(b) Has dissolved exactly as much solute as is theoretically possible
(c) Is unstable because it has dissolved more solute than would be expected
(d) None of the above

24. Which of the following is water tube boiler?
(a) Cochran (b) Lancashire (c) Locomotive (d) Babcock and Wilcox

25. ------------ mounting on the boiler regulates the steam pressure in boiler.
(a) Fudge plug (b) Pressure gauge (c) Safety valve (d) Feed check valve

26. ---------- is provided in boiler to increase the steam temperature above saturation temperature.
(a) Air pre heater (b) Super heater (c) Economizer (b) All of the above

27. Energy possessed by a system due to its molecular arrangement and motion of molecules is------- energy.
(a) Kinetic (b) Potential (c) Internal (d) Thermal

28. If two systems are in thermal equilibrium with a third system, they are also in thermal equilibrium with each other is the statement of-------------
(a) First law of thermodynamics (b) Second law of thermodynamics
(c) Zeroth law of thermodynamics (d) None of the above

29. ----------------- is the dry cleaning method.
(a) Soaking (b) Spraying
(c) Ultrasonic cleaning (d) Aspiration

30. Forces utilized in ball mill are ---------------
(a) Shear and impact (b) Impact and compression
(c) Compression and shear (d) None of the above

31. The space between individual wires of a wire mash screen is known as------------
(a) Mesh number (b) Screen aperture (c) Screen interval (d) Diameter of screen

32. --------is an operation in which uniform combination of two or more components is achieved.
(a) Centrifugation (b) Mixing (c) Filtration (d) Milling

33. --------------- is a unique example of emulsion technology.
(a) Butter making (b) Ice cream preparation
(c) Cream separation (d) All of the above

34. According to the laws of thermodynamics which of the following can be recycled?
(a) Both matter and energy (b) Neither matter nor energy
(c) Matter but not energy (d) Energy but not matter

35. What is stated by the First Law of Thermodynamics?
(a) Energy can be created or destroyed by physical processes
(b) Entropy always increases
(c) Energy cannot be created or destroyed
(d) The organization of the universe is steadily increasing

36. Which equipment is used for inclined or horizontal continuous movement?
(a) Conveyor (b) Elevator (c) Cranes (d) None of the above

37. Nylons are made for---------------
(a) Dibasic acids (b) Diamines (c) Both a and b (d) None of the above

38. When different plastics are combined to get certain desirable properties are known as---------
(a) Monomers (b) Plasticizers (c) Homo polymers (d) Copolymers

39. Which of the following is a fragile packaging material?
(a) Glass (b) Plastic (c) Paper (d) Metal

40. Which of the following is transportation hazard?
(a) Stacking (b) Dampness
(c) Impact (d) Temperature and relative humidity

41. -------------- package does not have definite shape.
(a) Flexible (b) Rigid (c) Semi rigid (d) All of the above

42. Gases used in modified atmospheric packaging are-----------
(a) O_2, SO_2, CO_2 (b) O_2, CO_2, N_2 (c) CO_2, N_2, NH_3 (d) SO_2, NH_3, CO_2

43. ------------- paper is brown in colour.
(a) Glassine (b) Sulphited (c) Tissue (d) Kraft

44. Tetra pack are made up of---------------
(a) Paper, aluminum, polyethylene (b) Polypropylene, paper, polyethylene
(c) Paper, aluminum, polystyrene (d) Polyethylene, aluminum, polystyrene

45. When two films are laminated by using some adhesive is known as------------
(a) Thermo lamination (b) Wet lamination
(c) Extrusion lamination (d) None of the above

46. The commonly used filter aid is------------
(a) Cellulose acetate (b) Solid cake (c) Kieselguhr (d) Glass fibers

47. In expression, the separation is on the basis of --------------
(a) Compressive force (b) Centrifugal force
(c) Shear force (d) Hydraulic force

48. Disc bowl separator is used for--------------
(a) Separation of cream from milk (b) Refining of animal fats
(c) Clarification of fruit juice (d) All of the above

49. Basic filter medium in edge filter is-------------
(a) Filter leaf (b) Filter pile (c) Cylindrical tube (d) Filter plate

50. --------------- fuels are characterized by high sulphur and ash contents.
(a) Solid (b) Liquid (c) Gases (d) All of the above

51. ------------ is not a primary source of heat energy.
(a) Solid fuel (b) Liquid fuel (c) Gases (d) Electricity

52. Which is the indirect heating method?
(a) Dielectric heating (b) Microwave heating
(c) Resistance heating (d) None of the above

53. Which of the following is not a membrane separation process?
(a) Reverse osmosis (b) Electro dialysis (c) Ultrafiltration (d) Distillation

54. CO_2 is -------------- times heavier than air.
(a) 10 (b) 7.3 (c) 3.53 (d) 1.53

55. Who has developed hand operated refrigeration system?
(a) John Gorrie (b) Jacob Perkins (c) Jams Harrison (d) Alexander Krik

56. A refrigerating machine of 20 ton rating a capable of cooling at a rate of--------------Btu/min.
(a) 4000 (b) 200 (c) 400 (d) 2000

57. The application of refrigeration includes --------------
(a) Ice manufacturing (b) Preservation
(c) Freezing (d) All of the above

58. Which of the following is a secondary refrigerant?
(a) Ammonia (b) Water (c) CO_2 (d) R12

59. Azeotrope R500 contains--------------
(a) 73.8% R12 and 26.2% R152a (b) 75% R22 and 25% R12
(c) 73.8% R22 and 26.2% R12 (d) 75% R12 and 25% R1529

60. The molecular formula of R12 is-------------
(a) CCl_2F_2 (b) $C_2Cl_2F_4$ (c) CCl_3F (d) $C_2F_3Cl_3$

61. --------------- is the representative of running cost.
(a) Compression ratio (b) Critical temperature
(c) Coefficient of performance (d) Condenser pressure

62. Which of the following is non toxic and non flammable refrigerant?
(a) NH_3 (b) SO_2 (c) R12 (d) Butane

63. ---------------- is used for detection of minor leakage of halocarbons.
(a) Electronic leak detector (b) Soap solution
(c) Sulphur candle (d) Ammonia swab

64. Which of the following is a treatment of water for making ice?
(a) Removal of core water (b) Removal of bubbles
(c) Agitation (d) All of the above
65. ----------------is considered as the heart of refrigeration system.
(a) Evaporation (b) Compressor (c) Condenser (d) Expansion valve
66. Which type of condenser is more preferable where acute water shortage exists?
(a) Double pipe condenser (b) Evaporative condenser
(c) Shell and tube condenser (d) None of the above
67. --------------- evaporator is always kept filled with liquid refrigerant.
(a) Plate (b) Fin and tube (c) Flooded (d) Dry expansion
68. Which freezer is suitable for liquid food?
(a) Scraped surface (b) Plate type (c) Immersion type (d) Air plate type
69. Which equation is used for predicting freezing time for food?
(a) Planck's equation (b) Pham's equation
(a) Both a and b (b) None of the above
70. Air conditioning is the simultaneous control of ------------
(a) Temperature (b) Humidity (a) Air motion (d) All of the above
71. ----------------- compressor consists of high speed impeller.
(a) Screw type (b) Reciprocating (c) Centrifugal (d) Scroll type
72. The ---------------- panel is a symbolic representation of entire process of system.
(a) Graphic (b) Serigraphic (c) Non-graphic (d) None of the above
73. The circuit with thermocouple is called as------------
(a) Wheatstone bridge (b) Closed circuit
(c) Potentiometer circuit (d) Feed back
74. The first requirement for pneumatic transmission is--------------
(a) Constant water supply (b) Constant air supply
(c) Constant force (d) Constant displacement
75. Thermometer used for measurement of temperature without physical contact is ------------
(a) Thermocouple (b) Resistance thermometer
(c) Pyrometer (d) Bimetallic thermometer
76. Which of the following is not a pressure element?
(a) Bourdon tube (b) Diaphragm (c) Bellows (d) Manometer
77. ---------------- scale is sometimes referred to as the absolute centigrade scale.
(a) Kelvin (b) Rankine (c) Fahrenheit (d) Reaumur
78. In gravity crusher----------------
(a) Movable jaw is pivoted at top (b) Maximum movement at bottom
(c) Both a and b (d) None of the above

79. In ball mill, at which speed there will be centrifugation?
(a) Operating (b) Normal (c) Critical (d) Centrifugation

80. Electrostatic separator makes use of difference in--------------
(a) Magnetic properties (b) Densities
(c) Electrical properties (d) Moisture content

81. For the flow of the fluid, difference in------------ is required.
(a) Pressure (b) Moisture content (c) Concentration (d) Force

82. Unit of specific gravity is-------------
(a) g/cm (b) kg/m (c) kg/cm (d) None of the above

83. ------------- solids are unfloatable.
(a) Hydrophilic (b) Hydrophobic (c) Both a and b (d) None of the above

84. The separation of solids from suspension in liquid by gravity is called as--------------
(a) Centrifugation (b) Sedimentation (c) Flocculation (d) None of the above

85. Axial flow, high speed impeller is nothing but a ----------------
(a) Propeller (b) Paddles (c) Turbines (d) Baffles

86. Which of the following blades are used for double arm kneader?
(a) Sigma (b) Double (c) Dispersion (d) All of the above

87. Fluids consist of--------------
(a) Liquids (b) Vapours (c) Gases (d) All of the above

88. The viscosity of ---------------- increases with temperature.
(a) Hydrogen (b) Milk (c) Water (d) Syrup

89. In case of fluid the ratio between shear stress and shear rate is------------
(a) Density (b) Viscosity (c) Specific gravity (d) Yield value

90. Monometer is used in------------
(a) Static condition of fluid (b) Dynamic condition of fluid
(c) Both a and b (d) None of the above

91. The density of monomeric fluid should be-------------- than that of the flowing fluid.
(a) Greater (b) Equal (c) Smaller (d) Any density

92. For the laminar flow Reynolds number should be less than------------
(a) 2100 (b) 4100 (c) 3100 (d) 5100

93. The ratio of inertia force to viscous force is called as-------------
(a) Nusselt Number (b) Prandtl Number (c) Reynolds Number (d) Newton Number

94. Tendency of fluids to segregate towards the walls and to flow along the walls is termed as--
(a) Capillary movement (b) Stream line flow
(c) Channeling (d) Diffusion

95. Metal pieces employeed to extend or increase the heat transfer surface are called as-----------
(a) Craters (b) Baffles (c) Fins (d) Separators

96. In boiling point diagram, the saturated vapour curve is termed as-------------point.
(a) Triple (b) Dew (c) Boiling (d) Saturation

97. The moisture content in excess of equilibrium moisture content is termed as-----------
(a) Saturated moisture (b) Specific moisture
(c) Free moisture (d) None of the above

98. Which of the following is variable area meter?
(a) Venturimeter (b) Orifice meter (c) Rotameter (d) All of the above

99. The pressure drop through globe valve is much---------- than through gate valve.
(a) Greater (b) Equal (c) Lesser (d) Either greater or lower

100. In distillation ---------------- takes place in both the direction.
(a) Heat transfer (b) Evaporation (c) Mass transfer (d) None of the above

101. Relative volatility is the measure of separation by --------------
(a) Gel filtration (b) Osmosis (c) Distillation (d) None of the above

102. Separation by distillation is not possible when relative volatility is equal to --------------
(a) Zero (b) Two (c) One (d) Four

103. Heat can be transmitted by --------------- mode across absolute vacuum.
(a) Conduction (b) Radiation (c) Convection (d) None of the above

104. Reciprocating compressors are --------------- speed machines.
(a) Slow (b) Medium (c) High (d) Critical

105. Pressure is always in a--------------
(a) Horizontal stress (b) Vertical strain (c) Normal stress (d) Horizontal strain

106. Force per unit area is ---------------
(a) Pressure (b) Stress (c) Both a and b (d) None of the above

107. Unit of viscosity is ---------------
(a) Pascal (b) Pascal meter (c) Pascal second (d) None of the above

108. If added sugars appear first or second on the food label list for a packaged food, it means----
(a) The list is alphabetical (b) It's high in sugar
(c) It's low in sugar (d) The position on the list is arbitrary

109. Wet bulb temperature is the measure of-------------
(a) Sensible heat of air (b) Latent heat of air
(c) Total heat of air (d) Specific heat of water

110. At super critical temperature, carbon dioxide acts as a -------------
(a) High density liquid (b) Tightly packed solid
(c) Low density liquid (d) Fast moving gas

111. Natural convection is characterized by -------------
(a) Grash of number (b) Reynolds number (c) Peclet number (d) Prandtl number
112. The entropy is------------
(a) Intensive property (b) Extensive property
(c) Derived property (d) None of the above
113. The thermal conductivity is minimum for ------------
(a) Silver (b) Aluminum
(c) Chrome-Nickel-Steel (d) Carbon steel
114. The angle formed by pouring a power as a heap on flat surface is known as------------
(a) Contact angle (b) Angle of repose (c) Angle of nip (d) Critical angle
115. ------------ is the heart of biogas plant.
(a) Inlet pipe (b) Mixing tank (c) Out let pipe (d) Digester
116. The colorific value of natural gas is -------------KJ/Kg.
(a) 50 (b) 60 (c) 70 (d) 80
117. Which of the following gaseous fuels is likely to have the highest gross calorific value on per m^3 basis?
(a) Sewage gas (b) Producer gas (c) LPG (d) Natural gas
118. Boiler mainly consists of ---------------
(a) Fire box (b) Heat exchanger (c) Flue (d) All of the above
119. Fire tube boilers are also called-------------boiler.
(a) High mass (b) High pressure (c) Low mass (d) Low pressure
120. Which of the following is the safety valve used in boiler?
(a) Spring loaded valve (b) Feed check valve
(c) Blow off valve (d) All of the above
121. Who has got the first British patent for artificial refrigeration?
(a) Thomas Harrison and John Long (b) Dr. John Gorrie
(c) Jacob Perking (d) Dr. Alexander Kirk
122. When a system undergoes a series of processes and return to its initial state is called as------
(a) Equilibrium (b) Work (c) Cycle (d) None of the above
123. The energy by the virtue of atomic and molecular configuration of a system is called as----------- energy.
(a) Heat (b) Internal (c) Molecular (d) Potential
124. Total heat content of the substance is it's-------------
(a) Latent heat (b) Enthalpy (c) Specific heat (d) Sensible heat
125. 1 BTU is approximately equal to--------------
(a) 1 J (b) 1 calorie (c) 1 kJ (d) 1kcalorie

126. Which of the following is/are the major component/s of a vapour compression system?
(a) Atomizer (b) Homogenizer (c) Evaporator (d) All of the above

127. Which of the following is the primary refrigerant?
(a) NaCl (b) Water (c) CaCl (d) Ammonia

128. Mixture of two or more refrigerant acting as a single refrigerant is called as--------------
(a) Enatiotrope (b) Azeotrope (c) Polytrope (d) Zoetrope

129. In blast freezing------------- is used.
(a) Liquid refrigerant (b) Cold air
(c) Plate heat exchanger (d) None of the above

130. Which of the following are/is as throttling device/s?
(a) Capillary tube (b) High side float valve
(c) Thermostatic (d) All of the above

131. Coefficient of performance of a refrigerant is the representative of--------------
(a) Running cost (b) Solubility in oil (c) Efficiency (d) Freezing point

132. Ammonia is being represented by-------------
(a) R 717 (b) R 790 (c) R 723 (d) R 700

133. Epoxy resins come under the category of --------------
(a) Thermoplastic (b) Oil soluble or oil modified
(c) Thermosetting (d) Protein substances

134. In a food industry, break-even point occurs when-----------------
(a) The annual rate of production equal to assigned value
(b) The total annual product cost equal to total annual sales
(c) The annual profit equals the expected value
(d) The annual sales equal the fixed costs

135. Solvent used in extractive distillation------------------
(a) Is of low volatility
(b) Forms a low boiling azeotrope
(c) Forms a high boiling azeotrope
(d) Does not alter the relative volatility of the original components

136. ------------- is stage wise gas-liquid contactor.
(a) Packed column (b) Plate column (c) Vertical column (d) Open column

137. At -------------- moisture content constant rate period ends and falling rate period starts.
(a) Critical (b) 90% (c) Specific (d) Initial

138. In the diverging section of venture meter, large proportion of kinetic energy is converted back to ----------------energy.
(a) Kinetic (b) Potential (c) Pressure (d) Heat

139. Which of the following pumps is used for boiler feed water applications?
(a) Gear pump (b) Metering pump (c) Plunger pump (d) All of the above

140. In case of blower, the---------------- ratio is below 3 and in case of compressors, it may be as high as 10 or more.
(a) Cavitation (b) Decavitation (c) Compression (d) Expansion

141. SI unit of overall heat transfer coefficient is--------------
(a) $W/(m^2K)$ (b) Wm^2K (c) m^2K/W (d) WK/m^2

142. The process of breaking down of solid material through the application of mechanical force is called as---------------
(a) Centrifugation (b) Filtering (c) Size reduction (d) Agitation

143. Thermal conductance is reciprocal of--------------
(a) Temperature difference (b) Cross sectional area of the conductor
(c) Thermal conductivity (d) Thermal resistance

144. Polished surface have----------------emissivity.
(a) High (b) Moderate (c) Low (d) Zero

145. Which law in mass transfer is similar to Fourier`s law for the heat transfer?
(a) Coulomb`s law (b) Stephan`s law (c) Fick`s law (d) Curie`s law

146. Temperature at the surface of the water body is the measure of---------------
(a) Dry bulb thermometer (b) Bulb temperature
(c) Mercury bulb thermometer (d) Wet bulb thermometer

147. Crushing efficiency is the ratio of --------------
(a) Surface energy created by crushing to the energy absorbed by the solid
(b) Energy absorbed by the solid to that fed to the machine
(c) Energy fed to the machine to the surface energy created by crushing
(d) Energy absorbed by the solid to the surface energy created by crushing

148. Rittinger's crushing law states that --------------
(a) Work required to form a particle of any size is proportional to the square of the surface to volume ratio of the product
(b) Work required to form a particle of a particular size is proportional to the square root of the surface to volume ratio of the product
(c) Work required in crushing is proportional to the new surface created
(d) For a given machine and feed, crushing efficiency is dependent on the size of the feed and product

149. Bond's crushing law--------------
(a) Calls for relatively less energy for the smaller product particles, than does the Rittinger's law
(b) Is less realistic in estimating the power requirements of commercial crushers
(c) States that the work required to form particle of any size from very large feed is proportional to the square root of the volume to surface ratio of the product
(d) States that the work required for the crushing is proportional the new surface created

150. Work index is defined as the ---------------
   (a) Gross energy (kWh/ton of feed) needed to reduce very large feed to such a size that 80% of the product passes through a 100 micron screen
   (b) Energy needed to crush one tone of feed to 200 microns
   (c) Energy (kWh/ton of feed) needed to crush small feed to such a size that 80% of the product passes a 200 mesh screen
   (d) Energy needed to crush one ton of feed to 100 microns
151. Fibrous material are broken by a---------------
   (a) Roll crusher (b) Squirrel cage disintegrator
   (c) Ball mill (d) Tube mill
152. Ribbon blenders are exclusively meant for--------------
   (a) Blending miscible liquids (b) Non-flowing powder and thin pastes
   (c) Batch mixing (d) Continuous mixing
153. Traces of solids are removed from liquid in a ---------------
   (a) Classifier (b) Clarifier (b) Sparkler filter (d) Rotary vacuum filter
154. Filter aid is used to---------------
   (a) Increase the rate of filtration (b) Decrease the pressure drop
   (c) Increase the porosity of the cake (d) Act as a support base for the septum
155. The most common filter aid is-------------
   (a) Diatomaceous earth (b) Calcium silicate
   (c) Sodium carbonate (d) Silica gel
156. Paddle agitator----------------
   (a) Is suitable for mixing low viscosity liquids
   (b) Produces axial flow
   (c) Moves at very high speed
   (d) None of the above
157. The material is crushed in a gravity crusher by the action of---------------
   (a) Impact (b) Attrition (c) Compression (d) Cutting
158. Mesh indicates the number of holes per--------------
   (a) Square inch (b) Linear inch (c) Square foot (d) Linear foot
159. Cyclones are used primarily for separating--------------
   (a) Solids (b) Solids from fluids (c) Liquids (d) Solids from solids
160. Maximum size reduction in a ball mill is done by which action?
   (a) Attrition (b) Compression (c) Impact (d) Cutting
161. Fourier's law applies to the heat transfer by-----------------
   (a) Convection (b) Radiation (c) Conduction (d) All of the above

162. Which of the following statcment is correct?
(a) Rate = Driving force × Resistance (b) Driving force = Rate × Resistance
(c) Resistance = Driving force × Rate (d) Rate = Resistance / Driving force

163. An insulator should have----------------
(a) Low thermal conductivity (b) High thermal conductivity
(c) Less resistance to heat flow (d) A porous structure

164. Which of the following has the highest thermal conductivity?
(a) Brick (b) Air (c) Water (d) Silver

165. It is possible to take out tube bundle outside of shell and clean it in case of--------------
(a) Fixed tube sheet exchanger (b) Double pipe exchanger
(c) Floating head exchanger (d) None of the above

166. Heat flux is the time rate of heat transfer per unit--------------
(a) Length (b) Area (c) Volume (d) None of the above

167. As per Kirchhoff's law, the ratio of the total radiating power to the absorptivity of a body depends on the--------------
(a) Temperature of the body only (b) Wavelength of monochromatic radiation
(c) Both a and b (d) Nature of material of body

168. Heat sensitive material can be concentrated in an evaporator by employing-------------
(a) Vacuum (b) High pressure
(c) High resistance timer (d) None of the above

169. Fruit juices can be concentrated in a --------------- evaporator.
(a) Long tube (b) High pressure (c) Falling film (d) None of the above

170. Reason for operating an evaporator in multiple effects is to secure----------------
(a) Increased steam economy (b) Decreased steam consumption
(c) Both a and b (d) Increased capacity

171. Baffles in the shell side of a shell and tube heat exchanger--------------
(a) Increases the cross section of the shell side liquid
(b) Force the liquid to flow parallel to the bank
(c) Increases the shell side heat transfer co-efficient
(d) Decreases the shell side heat transfer co-efficient

172. Steam trap is used to --------------
(a) Condense the steam flowing in the pipeline
(b) Remove water resulting from partial condensation of steam
(c) Stop the supply of steam
(d) None of the above

173. In SI units, thermal conductivity is expressed in--------------
(a) Watt/m.K (b) Watt /m^3. K (c) Watt/m^2.K (d) Watt /m^4.K

174. Thermal diffusivity is most important in heat transfer by---------------

(a) Conduction (b) Radiation (c) Condensation (d) Convection

175. The ratio of the total emissive power to the absorptivity for all bodies is same at thermal equilibrium. This is ------------ law.

(a) Kirchhoff's (b) Planck's

(c) Wien's displacement (d) Stefan- Boltzman

176. Mass transfer coefficient is defined as---------------

(a) Flux = coefficient / concentration difference

(b) Coefficient = flux /concentration difference

(c) Flux = concentration difference / coefficient

(d) None of the above

177. Humidification involves mass transfer between a pure liquid phase and a fixed gas which is------

(a) Insoluble in the liquid (b) Soluble in the liquid

(c) Non- ideal in nature (d) Soluble in mixed temperature

178. Steady state temperature reached by a small amount of liquid evaporating into a large amount of unsaturated vapour gas mixture is called --------------temperature.

(a) Dry bulb (b) Wet bulb (c) Dew point (d) Adiabatic saturation

179. Dry bulb temperature of the gas is --------------- the wet bulb temperature.

(a) Less than (b) More than (c) Equal to (d) None of the above

180. In distillation, overhead product contains---------------

(a) Only one component (b) Two components

(c) Any number of components (d) Only saturated liquid

181. Flash distillation is--------------

(a) Same as differential distillation

(b) Used for multi component systems like crude refining

(c) Same as simple distillation

(d) Most useful for handling binary systems

182. Relative volatility does not change appreciably with the change in-------------

(a) Temperature (b) Vapour pressure of either component

(c) Total pressure (d) None of the above

183. In rectifying section of a continuous distillation column, the--------------

(a) Vapour is enriched with low boilers (b) Vapour is enriched with high boilers

(c) Liquid is stripped of high boilers (d) None of the above

184. In stripping section of continuous distillation column, the--------------

(a) Liquid is stripped of high boiler (b) Liquid is enriched with high boiler

(c) Vapour is stripped of low boiler (d) None of the above

185. Heat sensitive materials with very high content heat of vaporization may be economically separated using-------------
(a) Liquid extraction (b) Distillation (c) Evaporation (d) Absorption

186. A fluid is the one, which---------------
(a) Cannot remain at rest under the action of shear force
(b) Continuously expands till it fills any container
(c) Is incompressible
(d) Permanently resists distortion

187. In an incompressible fluid, the density is---------------
(a) Greatly affected by moderate changes in pressure
(b) Greatly affected only by moderate changes in temperature
(c) Can not affect with moderate changes in temperature and pressure
(d) Sensible to changes in both temperature and pressure

188. Newton's Law of Viscosity relates the ---------------
(a) Shear stress and velocity
(b) Velocity gradient and pressure intensity
(c) Shear stress and rate of angular deformation in a fluid
(d) Pressure gradient and rate of angular deformation

189. An ideal fluid is--------------
(a) Frictionless and incompressible (b) One which obeys Newton's Law of viscosity
(c) Highly viscous (d) None of the above

190. Priming is needed in a-------------------- pump.
(a) Reciprocation b. Gear (c) Centrifugal (d) Diaphragm

191. Which of the following is Newtonian fluid?
(a) Rubber latex (b) Sewage sludge (c) Quick sand (d) Non colloidal solution

192. A centrifugal pump is called a turbine pump, if it have a----------------
(a) Turbine type Impeller (b) Vanned diffusion casing
(c) Rotating vanned volute (d) None of the above

193. The point at which solid, liquid and gas phases co-exist, is known as-----------point.
(a) Freezing (b) Triple (c) Boiling (d) Boyle

194. Second law of thermodynamics is concerned with the----------------
(a) Amount of energy transferred (b) Direction of energy transferred
(c) Irreversible process only (d) Non cyclic process only

195. For an ideal gas, the enthalpy--------------
(a) Increases with rise in pressure (b) Decreases with rise in pressure
(c) Is independent of pressure (d) Is a path function

196. High excess air in combustion of fuel result in----------------
(a) Increased fuel combustion (b) Incomplete combustion
(c) Smoky flame (d) None of the above

197. Thermocouple is suitable for measuring---------------- temperature.
(a) Liquid (b) Very high (c) Very low (d) Both high and low

198. Which of following is desirable characteristic of an instrument?
(a) High drift (b) High fidelity (c) High measuring lag (d) Poor reproducibility

199. Psychrometer determines the------------------
(a) Humidity of gases (b) Moisture content of solids
(c) Water of crystallization (d) Hygroscopic nature of solids

200. Trickling filters are used for clarification of --------------
(a) Syrups (b) Wine (c) Sewage (d) Fruit juices

201. Flapper nozzle is ----------------- controller.
(a) Pneumatic (b) Hydraulic (c) Electronic (d) None of the above

202. In a ------------------riveted joint, the rivets in the various rows are opposite to each other.
(a) Chain (b) Zigzag (c) Diamond (d) None of the above

203. The thermal stress in metallic bar does not depend upon the----------------
(a) Change in temperature (b) Cross sectional area
(c) Both a and b (d) Neither a nor b

204. In ------------------- process industry product is processed in lots.
(a) Intermittent (b) Continuous (c) Repetitive (d) Cottage

205. The overall design of an enterprise is-----------------
(a) Plant design (b) Plant layout (c) Plant location (d) Product design

206. Steam is preferred to be used as a heating medium in heat exchange because of its-----------
(a) Low cost (b) High latent heat
(c) Non-corrosive condensate (d) High film-coefficient

207. Corrosion resistance of steel is increased by addition of---------------
(a) Phosphorous and tungsten (b) Chromium and nickel
(c) Lead and vanadium (d) Molybdenum and tungsten

208. In a ball mill, balls are usually made up of ----------------
(a) Cast iron (b) Steel (c) Stainless steel (d) Bronze

209. Glass is chemically ------------------- in nature.
(a) Reactive (b) Inert (c) Semi-reactive (d) None of the above

210. Packaging material should be------------------ to retain desirable properties of food products.
(a) Permeable (b) Impermeable (c) Semi permeable (d) None of the above

211. In paper manufacturing ------------------in wood pulp is removed by washing.
(a) Fibers (b) Lignin (c) Moisture (d) None of the above

212. PET is manufactured by ------------------ method.
(a) Reforming
(b) Stretch blow molding
(c) Blow molding
(d) Injection molding

213. ---------------- offers superior mechanical strength.
(a) Glass (b) Metal container (c) Paper (d) Plastic

214. The rigid metal containers are sterilized by using ---------------
(a) Hot water
(b) Super heated steam
(c) H_2O_2
(d) None of the above

215. When the mill is centrifuging----------------
(a) Maximum grinding takes place
(b) Little or no grinding takes place
(c) Normal grinding takes place
(d) None of the above

216. Screening means separation of the particle according to ------------- alone
(a) Size (b) Shape (c) Moisture content (d) Magnetic property

217. Rate of filtration is---------------
(a) Maximum at early stage
(b) Maximum at middle stage
(c) Maximum at final stage
(d) Uniform throughout the filtration process

218. The vacuum filters are limited to maximum filtering pressure of---------------
(a) One bar (b) Two bar (c) Three bar (d) Five bar

219. Automated filter is-------------
(a) Leaf filter
(b) Stationary drum filter
(c) Rotary drum filter
(d) Plate and frame filter press

220. Baffles in mixing tanks are provided to------------
(a) Improve the rate of mixing
(b) Minimize vertex formation
(c) Both a and b
(d) Neither a nor b

221. The equipment where sedimentation process is being carried out is called------------
(a) Thickeners (b) Centrifuge (c) Sedimenter (d) Propeller

222. Manometer is used in---------------
(a) Static condition of fluid
(b) Dynamic condition of fluid
(c) Neither a and b
(d) In both a and b

223. In crystallization from solution, simultaneous---------------- and mass transfer takes place.
(a) Heat (b) Vibration (c) Pressure (d) Temperature

224. Vacuum crystallizer is used for------------
(a) High viscous material
(b) Heat sensitive material
(c) Fast crystallization
(d) All of the above

225. An zoetrope is a---------------
(a) Type of isotope of uranium used for irradiation of meat
(b) Liquid mixture with vapour of same composition as that of liquid
(c) Type of heat exchanger where coolant used is liquefied gas
(d) Chemical compound used for the cleaning in place

226. The centrifugal pump depends upon the --------------- of a fluid to develop pressure.
(a) Viscosity (b) Energy (c) Inertia (d) Density

227. The head of centrifugal pump --------------- continuously as the capacity is decreased.
(a) Decreases (b) Increases (c) Becomes less (d) Becomes constant

228. Reciprocating compressors are -------------- speed machines.
(a) Slow (b) High (c) Medium (d) Critical

229. Heat transfer from hot fluid to cold fluid in heat exchange equipment takes place by conduction and ---------------- mode.
(a) Radiation (b) Convection (c) Both a and b (d) Neither a nor b

230. Which of the following is dry cleaning methods used for the cleaning of the raw material?
(a) Brushing (b) Flaming (c) Flotation (d) Spraying

231. In ultrasonic cleaning--------------- waves are used.
(a) Electric (b) Sound (c) Magnetic (d) All of the above

232. Economy of single effect evaporator is---------------
(a) Always more than one (b) Always less than one
(c) Always one (d) Generally 100

233. In hammer mill, the force involved is----------------
(a) Impact (b) Attrition (c) Crushing (d) Both a and b

234. Wet milling tends to produce ---------------- product.
(a) Coarser (b) Medium size
(c) Finer (d) There is no size reduction

235. Cutting of food material into cubes is called as---------------
(a) Slicing (b) Milling (c) Dicing (d) Crystallization

236. Mess number is number of opening of wire per-------------
(a) Linear inch (b) Linear centimeter (c) Linear millimeter (d) Linear meter

237. Vertical screw mixers are used for mixing of----------------
(a) High viscous liquids (b) Low viscous liquids
(c) Moderate viscous liquids (d) Dry solids

238. Which of the following is used as clarifier?
(a) Conduct angle bowl (b) Disc bowl centrifuge
(c) Tabular bowl centrifuge (d) Cylinder bowl centrifuge

239. Desludging involves --------------- force.
(a) Gravitational (b) Centrifugal (c) Magnetic (d) Electrical

240. Drying by heated air is called as---------------
(a) Conduction (b) Convection (c) Radiation (d) All of the above

241. Vapour pressure is the function of---------------
(a) Pressure (b) Temperature (c) Both a and b (d) Neither a nor b

242. Partial pressure is the function of ---------------
(a) Pressure (b) Number of molecules of water
(c) Both a and b (d) Neither a nor b

243. Dew point is the temperature at which the---------------occurs.
(a) Boiling (b) Evaporation (c) Condensation (d) Freezing

244. Evaporation occurs -------------------
(a) From throughout the mass
(b) From the surface
(c) From bottom
(d) Either of the above three depending in the direction of heat

245. Wet bulb temperature is the measure of---------------
(a) Sensible heat of air (b) Total heat of air
(c) Latent heat of water (d) Specific heat of water

246. In single effect evaporator, the economy is-----------------
(a) Zero (b) Less than one (c) One (d) More than one

247. In leaching, there is separation of solid from ---------------- by a solvent.
(a) Solid (b) Liquid (c) Gas (d) All of the above

248. Diffusion is a------------
(a) Molecular phenomenon
(b) Biochemical phenomenon
(c) Phenomena where there is no relative movement of molecule
(d) Electromagnetic process

249. Calorific value of biogas (60% CH_4) is -------------Kcal/m^3.
(a) 5713 (b) 4713 (c) 4500 (d) 3713

250. Toothpaste is a----------------
(a) Bingham plastic (b) Pseudo plastic (c) Newtonian fluid (d) Dilatants

251. Mixing of two fluids is a -------------
(a) Reversible process (b) Irreversible process
(c) Isothermal process (d) None of the above

252. Entropy is-------------
(a) Intensive property (b) Derived property
(c) Extensive property (d) None of the above

253. Heat transfer occurs by natural convection because change in temperature causes differences in -------------
(a) Viscosity (b) Density
(c) Thermal conductivity (d) Heat capacity

254. Grash of Number is associated with-----------
(a) Buoyancy effects (b) Free convection
(c) Forced convection (d) High temperature difference

255. Heat transfer coefficient in a helical coil compared to that in straight pipe is----------
(a) Lower (b) Higher (c) Same (d) None of the above

256. If the temperature of atmosphere increases at constant absolute humidity, the percentage saturation would -------------
(a) Decrease (b) Increase (c) Remain constant (d) None of the above

257. If the temperature of atmosphere increases at constant absolute humidity, the wet bulb temperature would------------
(a) Decrease (b) Increase (c) Remain constant (d) None of the above

258. When an exothermic reversible reaction is conducted adiabatically, the rate of reaction-------
(a) Continuously increases (b) Continuously decreases
(c) Passes through the maximum (d) Passes through the minimum

259. The most widely used coagulant for removing suspended impurities from water is-----------
(a) Bleaching powder (b) Chlorine
(c) Calcium sulphate (d) Alum

260. Which of the following gaseous fuels is likely to have the highest gross calorific value on per m^3 basis?
(a) Sewage gas (b) LPG (b) Producer gas (d) Natural gas

261. Which of the following fuels have highest gross calorific value?
(a) Diesel oil (b) Heavy fuel oil (c) Lignite coal (d) Anthracite coal

262. Fire tube boilers are also called-------------
(a) High mass boiler (b) Low mass boiler
(c) High pressure boiler (d) Low pressure boiler

263. Which of the following is/are the boiler mountings?
(a) Water level indicator (b) Economizer
(c) Feed pump (d) All of the above

264. Which of the following is the safety valve used in boiler?
(a) Spring loaded valve (b) Blow off valve
(c) Feed check valve (d) All of the above

265. Which of the following pumps have pulse setting discharge flow?
(a) Centrifugal pump (b) Rotary pump
(c) Reciprocating pump (d) All of the above

266. A rotameter measures---------------
(a) Flow rate of fluid (b) Speed of turbines
(c) Viscosity of liquid (d) None of the above

267. Measure of randomness of a system is called-------------
(a) Enthalpy (b) Entropy (c) Energy (d) Effectiveness.

268. Which of the following is/are the measure component/components of vapour compression system?
(a) Atomizer (b) Evaporator (c) Homogenizer (d) All of the above

269. First law of thermodynamics relates with the conversion of---------------
(a) Heat (b) Energy (c) Momentum (d) Temperature

270. Vapour compression system is-----------------
(a) Refrigeration system (b) Evaporation system
(c) Freeze drying system (d) Fermentation

271. Compressor in a vapour compression system is a-------------
(a) Essential part (b) Optional part (c) Safety device (d) All of the above

272. Which of the following is primary refrigerant?
(a) NaCl (b) $CaCl_2$ (c) Water (d) Ammonia

273. Ton of refrigeration is -----------------
(a) Unit to express the capacity of a refrigerant plant
(b) Amount of ice formed in refrigerator during 24 hr.
(c) Amount of coal required to complete one refrigeration cycle
(d) Amount of water converted to ice in 1 hr. by refrigerator

274. A good refrigerant should have -------------
(a) Low boiling point (b) High specific volume
(c) Low conductivity (d) High coefficient of performance

275. One ton of refrigeration is equal to--------------
(a) 200 BTU/min. (b) 200 KJ/min. (c) 288000 BTU/hr. (d) All of the above

276. Which of the following refrigerant posses the environmental problem?
(a) Ammonia (b) Methane chloride (c) R12 (d) CFC

277. Total heat content of the substance is it's--------------
(a) Latent heat (b) Specific heat (c) Enthalpy (d) Sensible heat

278. Higher is the saturation temperature--------------
(a) Lower is the cooling effect
(b) Higher is the cooling effect
(c) Higher is the refrigerant pressure at the compressor outlet
(d) Lower is the refrigerant pressure at the compressor outlet

279. Which of the following is/are as the throttling device/s?
(a) Capillary tube (b) Thermostatic
(c) High side float valve (d) All of the above

280. Standard temperature and pressure refers to----------------
(a) 0 atm and 273 K (b) 1 atm and 273 K
(c) 101.325 kPa and 0 K (d) All of the above

281. The vapour pressure of a liquid increases when------------
(a) The temperature is raised (b) The temperature is lowered
(c) The pressure is lowered (d) None of the above

282. A calorimeter is used to------------
(a) Determine the heat of a reaction
(b) Determine the heat given off/absorbed during some process
(c) Store the heat from a chemical reaction
(d) None of the above

283. Disposable non-woven wrap provides-------------
(a) Effective bacterial barrier (b) Marginal bacterial barrier
(c) Reduction of costs for disposal (d) Both a and c

284. The double wrapping method that uses two wrappers in a single step is called--------------- wrapping.
(a) Sequential (b) Simultaneous (c) Double fold (d) Triangular

285. When arranging plastic/paper peel pouches in the sterilizer they should be positioned standing up right on edge with------------
(a) Plastic to plastic (b) Paper to plastic (c) Paper to paper (d) However they fit best

286. All are true about containers except---------------
(a) Most durable sterilization packaging (b) Provide optimum barrier to protect contents
(c) Compatible with all sterilization methods (d) Easy to use and most versatile

287. Sterilization packaging systems have an impact on---------------
(a) Instruments (b) Healthcare acquired infection
(c) Healthcare costs (d) All of the above

288. Now a day's most commonly used packaging material for RTS beverage is----------
(a) Glass bottles (b) Tetra packs (c) Polyethylene (d) Cellophane

289. In which process temperature remains constant------------

(a) Adiabatic (b) Isothermal (c) Isometric (d) Isobaric

290. Density of any biological material is given by------------

(a) Rate of change of velocity (b) Force per unit area

(c) Weight per unit volume (d) Mass per unit volume

291. Upper screen of the air screen cleaner is known as-------------

(a) Grading screen (b) Sorting screen (c) Scalping screen (d) None of the above

292. A semi permeable membrane allows the flow of-------------

(a) Solvent molecules (b) Solute molecules

(c) Both solute and solvent (d) None of the above

293. The energy unit, the Newton Metre is called------------

(a) Joule (b) Watt (c) British thermal unit (d) None of the above

294. Paddle agitator--------------

(a) Is suitable for mixing low viscosity liquids

(b) Produces axial flow

(c) Moves at very high speed

(d) None of the above

295. Second law of thermodynamics is concerned with the--------------

(a) Amount of energy transferred (b) Direction of energy transferred

(c) Irreversible process only (d) Non cyclic process only

296. The most widely used material for food packaging is--------------

(a) Aluminum (b) Paper (c) Glass (d) Steel

297. As a commercial packaging material, which is the most significant property of glass?

(a) Chemical inertness (b) Thermal stability

(c) Compressive strength (d) Impact resistance

298. Which one of the following gas is commonly used as refrigerant in refrigerator?

(a) Carbon dioxide (b) Nitrogen (c) Hydrogen (d) Freon

299. Indian Institute of Packaging is located at---------------

(a) Chennai (b) Mumbai (c) Delhi (d) Mysore

300. Z value in thermal processing is defined as----------------

(a) °F required to reduce the TDP ten fold

(b) °F required to reduce the TDP five fold

(c) °F required to reduce the TDP fifteen fold

(d) None of the above

301. The angle of repose increases with the material when moisture content ----------------.

(a) Increases (b) Decreases

(c) Increases or decreases (d) None of the above

302. Unit of viscosity is----------------

(a) Pascal (b) Pascal meter (c) Pascal second (d) None of the above

303. For a given height of chimney and given mean temperature of chimney gases, the draught produced----------

(a) Decreases with increase in outside air temperature

(b) Increases with increase in outside air temperature

(c) Remains same irrespective of outside air temperature

(d) May increase or decrease with increase in outside air temperature

304. Which of the following is not commercial freezing equipment?

(a) Air blast freezer (b) Plate freezer

(c) Fluidized bed freezer (d) Chest freezer

305. Which of the following is recent technique used for packaging of milk that offers economy?

(a) Cans (b) Glass bottles

(c) Plastic bottles (d) Flexible plastic pouches

306. The energy released during pre cooling of fresh fruits is known as---------------

(a) Field heat (b) Latent heat (c) Vital heat (d) None of the above

307. --------------- is the difference between dry bulb and wet bulb temperature.

(a) Dry bulb depression (b) Wet bulb depression

(c) Dew point depression (d) Degree of saturation

308. Heat enters through cracks of doors, windows, ventilators, walls etc. is called --------------

(a) Sun load (b) Infiltration load (c) Wall gain load (d) None of the above

309. The term $KA \div \Delta x$ is called as ---------------- of wall.

(a) Thermal resistance (b) Thermal conductance

(c) Heat flux density (d) Thermal loading

310. Mass transfer occurs due to gradient in---------------

(a) Temperature (b) Velocity (c) Concentration (d) None of the above

311. --------------causes very high rate of heat transfer and having high heat transfer coefficient due to bare areas.

(a) Film type condensation (b) Drop wise condensation

(c) Condensation boiling (d) None of the above

312. ------------------ is the ability of material to conduct and store the heat.

(a) Specific heat (b) Thermal conductivity

(c) Electrical conductivity (d) Thermal diffusivity

313. A screw conveyor is used up to a maximum inclination of -------------- from horizontal.

(a) 10° (b) 15° (c) 20° (d) 30°

314. For high speed bucket elevators with centrifugal discharge "A" is equal to --------------
(a) 1.0 (b) 1.0 to 1.4 (c) 1.5 to 3.0 (d) 3.5

315. The flow behaviour index (n) for --------------- fluid is greater than one.
(a) Newtonian (b) Pseudoplastic (c) Dilatent (d) Bingham plastic

316. The liquid component that passes through membrane during filtration is called as---------
(a) Feed slurry (b) Filtrate (c) Filter medium (d) Pre coat

317. The freezing point of food is --------------- that of pure water.
(a) Equal to (b) Lower than (c) Higher than (d) Twice than

318. Scientific calculation is based on moisture content on ---------------- weight basis.
(a) Wet (b) Dry (c) Dry and Wet (d) None of the above

319. ------------------- is suitable for handling powdery and sticky material.
(a) Belt conveyor (b) Bucket elevator
(c) Pneumatic conveyor (d) Screw conveyor

320. Weight sorter is not used for-------------
(a) Eggs (b) Apple (c) Paddy (d) Orange

321. The chemical used for determination of specific gravity of granular agricultural materials is---------
(a) Freon (b) Toluene (c) Hydrochloric acid (d) Sulphuric acid

322. In attrition mill, the size of food grains is reduced by-------------
(a) Shear and crushing (b) Impact and shear
(c) Impact and crushing (d) None of the above

323. To separate mustard seed from wheat ---------------- is most suitable separator.
(a) Specific gravity (b) Indented cylinder (c) Spiral (d) Pneumatic

324. Additions of salt to ice will ---------------- the temperature of the mixture.
(a) Increase (b) Decrease (c) Do not alter (d) None of the above

325. In attrition mill, overfeeding results in--------------
(a) Excessive heating (b) Reduces the effectiveness of the grinder
(c) Increases the capacity (d) All of the above

326. ---------------- indicates the relative proportions of the coarse, medium and fine particles.
(a) Fineness modulus (b) Uniformity index
(c) Average particle size (d) Effectiveness

327. ---------------- gives the relationship between energy consumed and size attained.
(a) Fick's law (b) Plank's law (c) Fourier's law (d) Bond's law

328. Specific gravity seed separator is used for grading the seed on the basis of-------------
(a) Same size but different sp. gravity (b) Same size and same sp. gravity
(c) Different size and different sp. gravity (d) Different size and same sp. gravity

329. The angle of repose increases with the material when moisture content--------------

(a) Increases (b) Decreases
(c) Increases or decreases (d) None of the above

330. Drying rate is --------------- of the relative humidity of air.

(a) Dependent (b) Independent (c) Both a and b (d) All of the above

331. Texture profile is measured by using----------------

(a) Viscometers (b) GC-MS (c) Textrometers (d) Both a and b

332. ------------ is the ratio of amount of water vapour actually present to the maximum amount of water vapour could hold at the same temperature.

(a) Humidity (b) Relative humidity (c) Absolute humidity (d) All of the above

333. The electromagnetic wave has an energy which is ----------------

(a) Inversely related to the wave length (b) Directly related to the wave length
(c) No relation with wave length (d) None of the above

334. In TDT curve-----------

(a) The D values are plotted on a logarithmic scale
(b) The corresponding temperature on a linear scale
(c) Both D values and temperature are on a linear scale
(d) Both a and b

335. The thermal death time curve is always a ------------

(a) Straight line (b) Sigmoid (c) U shape (d) None of the above

336. ---------------- yields maximum head rice and effectively separates the bran.

(a) Huller (b) Sheller (c) Huller cum sheller (d) Rubber-roller-rice mill

337. Porosity depends on----------------

(a) Shape (b) Dimensions (c) Roughness (d) All of the above

338. The differential speed ratio of break rolls of wheat mill is about------------

(a) 1:1 (b) 2:1 (c) 2.5:1 (d) 1:2

339. Grading is sorting on the basis of------------

(a) Quantity (b) Quality (c) Simple (d) None of the above

340. Rheology is the study of ---------------- properties of the dough.

(a) Elastic (b) Viscous (c) Viscoelastic (d) Both b and c

341. Loss of moisture by evaporation during milling is known as ---------------

(a) Milling drying (b) Milling loss (c) Moisture loss (d) All of the above

342. --------------- dryer is recommended for parboiled rice.

(a) Drum (b) Conveyor (c) Rotary (d) Tunnel

343. Vacuum packaging ---------- development of rancidity.

(a) Retards (b) Neither retards nor accelerates
(c) Accelerates (d) Sometimes retards while sometimes accelerates

344. The panel boards are ------------------

(a) Graphic (b) Non-graphic (c) Semigraphic (d) All of the above

345. Aluminums cans is ------------% barrier to WVIR.

(a) 95 (b) 98 (c) 100 (d) 97

346. --------------- are made from dibasic acids combined with diamines.

(a) Nylons (b) Copolymer (c) PVC (d) PET

347. ----------- containers are widely used for packing mineral water.

(a) PVC (b) PET (c) PP (d) PVA

348. Greases proof paper are used in packaging of ------------

(a) Fresh fruits and vegetables (b) Jam and jelly products

(c) Baked foods (d) Frozen foods

349. Combining two or more than two films into a single film is known as------------

(a) Kraft paper (b) Lamination (c) Restorable pouches (d) End folds wrapping

350. The packaging which helps to extend shelf life and improves quality, safety and sensory properties of packed food is known as ---------------

(a) Intelligent packaging (b) Active packaging

(c) NMBP packaging (d) Microbial packaging

351. ----------- is a carbon dioxide emitter.

(a) Ethanol (b) Ascorbic acid

(c) Sorbic acid (d) Potassium permanganate

352. --------------gas is used in modified atmosphere packaging of potato wafers.

(a) Carbon dioxide (b) Nitrogen (c) Oxygen (d) Carbon monoxide

353. Lacquers cans are used for packaging of ---------------products.

(a) Bakery products (b) Coloured fruit (c) Frozen fruits (d) Milk powder

354. Easy open end cans are widely used in-------------- packaging.

(a) Smart (b) Beverages and fruit juice

(c) Fruit pulp (d) Dried powder

355. The cold milk entering at 5^{0}C, after regeneration at 63^{0}C and final temperature of 71^{0}C would have an ------------% regeneration.

(a) 87.9 (b) 90.0 (c) 87.0 (d) 86.9

356. The unit of flow is ----------------

(a) Cumec/Cusec (b) Litres (c) Gallon (d) None of the above

357. The movement of moisture in falling rate regime is governed by--------------

(a) Osmosis (b) Permeability (c) Diffusion (d) Thermal conductivity

358. ----------- is useful to determine whether a product will gain or lose moisture under given set of temperature and relative humidity.

(a) Wet basis moisture content (b) Critical moisture content

(c) Equilibrium moisture content (d) Dry basis moisture content

359. The enthalpy of 1 Kg steam at a pressure of 8 bar and dryness fraction 0.8 will be -------- KJ/Kg if hf and hfg corresponding to 8 bar are 720.9 and 2046.5 KJ/Kg respectively.

(a) 2767.4 (b) 2358.1 (c) 1325.6 (d) None of the above

360. ----------------is the measure of the water content of air.

(a) Relative Humidity (b) Absolute Humidity
(c) Humidity (d) Equilibrium relative humidity

361. -------------- temperature is the measure of specific heat of water.

(a) Dry bulb (b) Wet bulb (c) Dew point (d) Absolute

362. --------------- temperature is the measure of sensible heat of air.

(a) Dry bulb (b) Dew bulb (c) Wet point (d) Absolute

363. --------- in mass transfer is analogus to Nusselt number in heat transfer.

(a) Grashof number (b) Rayleigh number (c) Reynold number (d) Sherwood number

364. Natural convection is characterized by-------------

(a) Grashof number (b) Sherwood number (c) Reynold number (d) Nusselt number

365. Stefan-Boltzmann law is applicable for heat transfer by---------------

(a) Conduction (b) Convection
(c) Conduction and Convection (d) Radiation

366. What does 1-4 shell and tube heat exchanger imply?

(a) 1 shell side pass and 4 tube side pass (b) 4 shell side pass and 1 tube side pass
(c) 4 shell side pass and 4 tube side pass (d) 1 shell side pass and 1 tube side pass

367. The specific heat of water is ----------------

(a) 2 kJ /kg0C (b) 4.386 kJ/kg0C (c) 4.186 kJ/kg0C (d) 3.186 kJ/kg0C

368. -------- is mechanical device designed to separate suspended particles from a gaseous or liquid steam.

(a) Magnetic separator (b) Screen separator
(c) Cyclone separator (d) All of the above

369. In batch process of expression - - - - ------press can be used for fruit juice expression.

(a) Screw press (b) Hydraulic ram (c) Magnetic press (d) None of the above

370. During refrigeration cycle, heat is rejected by the refrigerant in-------------

(a) Compressor (b) Capillary (c) Condenser (d) Evaporator

371. The rotating spindle apparatus is used to measure -------------- of liquid.

(a) Plasticity (b) Resistivity (c) Elasticity (d) Viscosity

372. The circuit used with the thermocouple is called -------------------

(a) Electrical circuit (b) Potentiometer circuit
(c) Electronic circuit (d) None of the above

373. The relative humidity is determined by knowing ----------------

(a) Wet bulb thermometer reading (b) Dry bulb thermometer reading

(c) Both a and b (d) None of the above

374. The thermometer measures the temperature - 40^0C, which is equal to ---------------

(a) -400^0F (b) -4^0F (c) -40^0F (d) None of the above

375. ------------------- instrument mentioned below is not manometer.

(a) U tube (b) Inclined tube (b) Well type (d) Pyrometer

376. ------------------- mentioned below is not temperature scale.

(a) ^{0} Kelvin (b) ^{0} Rankine (c) ^{0} Brix (d) ^{0} Centigrade

377. During refrigeration cycle, heat is absorbed by the refrigerant in-----------

(a) Throttling valve (b) Capillary (c) Condenser (d) Evaporator

378. The refrigerant R12 is called as-------------

(a) Monochlorotrifluro methane (b) Trichloromonofluro methane

(c) Dichlorodifluromethane (d) All of the above

379. -------- was the first refrigerant used by human being.

(a) Ammonia (b) Sulphurdioxide (c) Methyl chloride (d) Natural ice

380. The latent heat of evaporation of liquid nitrogen at -196^0C is ------------

(a) 80 kJ/kg (b) 335 kJ/kg (c) 2257 kJ/kg (d) 200 kJ/kg

381. Indirect liquid level measurement is done by ----------------

(a) Pressure gauge (b) Float (c) Bob and tape (d) Sight glass

382. ---------------is the temperature at which a given sample of the moist air will become saturated and deposit dew.

(a) Wet bulb temperature (b) Dew point temperature

(c) Dry bulb temperature (d) All of the above

383. ---------------- is the science that involves the study of chemical reactions and mechanisms.

(a) Dynamics (b) Statics (c) Kinetics (d) All of the above

384. The energy level that molecule must achieve to initiate a chemical reaction is called as ----------- energy.

(a) Kinetic (b) Potential (c) Store (d) Activation

385. The sight glass is the ---------------liquid level measurement device.

(a) Indirect (b) Pressure gauge (c) Direct (d) None of the above

386. The conversion of one form of energy into another is produced by a device called ---------

(a) Dynamometer (b) Thermocouple (c) Transducer (d) All of the above

387. Vapour-field thermometers can measure temperatures from ---------------

(a) -140^0C to $+315^0$C (b) -40^0C to $+40^0$C

(c) -40^0C to 400^0C (d) All of the above

388. Iron-constantan thermocouples has temperature range of ---------------
(a) -20^{0}C to 760^{0}C (b) 20^{0}C to 760^{0}C (c) -200^{0}C to -400^{0}C (d) None of the above

389. The actual vacuum in a condenser is equal to ------------
(a) Barometric pressure + actual pressure (b) Barometric pressure - actual pressure
(c) Gauge pressure + atmospheric pressure (d) Gauge pressure - atmospheric pressure

390. A good plant layout, in general, has the following advantages---------------
(a) Saving in floor space
(b) Better utilization of machine and man power and services
(c) Reduced material handling, thus saving in labour and cost, less production delays
(d) All of the above

391. Generally the small food processing plants have process form of layout, where bigger industries have ----------- layouts.
(a) Multi storey (b) Combined (c) Single level (d) Hybrid

392. The food industry generally needs electrical wiring of----------------
(a) 3 phase 4 wire (b) 3 phase 3 wire (c) Single phase supply (d) All of the above

393. The frequency of inspection of machines is affected by--------------
(a) Age of machine (b) Condition of machine
(c) Value of machine (d) All of the above

394. The technology of using instrument to measure and control the physical and chemical properties of materials is called--------------
(a) Material science (b) Statics (c) Dynamics (d) Instrumentation

395. The monitoring of control devices is accomplished through the use of -------------- signals.
(a) Alarm (b) Digital (c) Electrical (d) None of the above

396. Unit of measurement of noise is-------------
(a) Lux (b) Decibel (c) Faranhite (d) Joule

397. Resistance to flow in non-Newtonian fluid is called as -------------
(a) Viscosity (b) Consistency (c) Density (d) Specific gravity

398. The SI unit to express the energy value of food is---------------
(a) Kcal (b) Calorie (c) Kj (d) Joule

399. The gas is available at a pressure of about 10 cm of water column is -----------
(a) Janta Model (b) Deenbandhu model (c) KVIC Model (d) Batch Type Model

400. The amount of reflection of light at given wavelength is called as------------
(a) Chroma (b) Hue (c) Value (d) Colour

401. Viscosity of dilatants material ----------------- as shear rate increases.
(a) Decreases (b) Increases (c) Remains same (d) None of above

402. Gloss is measured by -----------
(a) Spectrophotometer (b) Goniophotometer
(c) Colorimeter (d) Munshell colour chart

403. --------------- is an instrument used for determining the area within closed curve.
(a) Succolometer (b) pH meter (c) Planimeter (d) Venire caliper

404. Instron texture measuring instrument is based on ------------- curve.
(a) Time– Force (b) Pressure– Time (c) Force– Pressure (d) None of the above

405. How does modified atmospheric packaging preserves food?
(a) It prevent enzyme activity (b) It alters the gases surrounding the food
(c) It prevents entry of oxygen to product (d) It limits water available for microbial growth

406. Density can be measured as---------------
(a) g/lit (b) m/sec (c) kg/ lit (d) None of the above

407. Zero energy cool chambers works on the principle of -----------------cooling.
(a) Hydro (b) Evaporative (c) Vacuum (d) Room

408. Solid CO_2 is also known as-------------
(a) Nitrogen (b) Carbon mono chloride
(c) Dry ice (d) Liquid state

409. In controlled atmospheric storage, the concentration of ---------- is low and concentration of CO_2 is high.
(a) Nitrogen (b) Hydrogen (c) Oxygen (d) Water

410. Rate of dehydration depend upon----------------
(a) Temperature (b) Surface area (c) Air velocity (d) All of the above

411. Field heat from harvested fruits and vegetables is removed by which technique?
(a) Irradiation (b) Pre cooling (c) Packing (d) Grading

412. The flow with unchanging velocity distribution is called as-------------flow.
(a) Potential (b) Steady (c) Unsteady (d) Fully developed

413. The velocity of the flowing liquid at the contact of pipeline is---------------
(a) Twice the average velocity (b) Twice the minimum velocity
(c) Half of the maximum velocity (d) Zero

414. In case of rectification, the portion of the condensed liquid that is returned to column is called as---------
(a) Condensate (b) Reflux (c) Efflux (d) None of the above

415. Which of the following fluids are expected to show the characteristics of laminar flow?
(a) High viscous (b) Moderate viscous
(c) Low viscous (d) Laminar flow does not depend on viscosity

416. Manometer is a simple device for measuring differences of-----------------
(a) Force (b) Pressure (c) Density (d) Temperature

417. Manometer is used in-------------

(a) Static condition of fluid (b) Dynamic condition of fluid

(c) Neither a and b (d) Both a and b

418. Which of the following packages is an example of aseptic packaging?

(a) Tetra pack drinking boxes (b) Paper bag

(c) Milk carton (d) Plastic bread bag

419. Which of the following property is not an intensive property of the system?

(a) Temperature (b) Pressure (c) Total energy (d) Density

420. Which of the following law provides basis for temperature measurement?

(a) First Law of Thermodynamic (b) Newton's Law

(c) Zeroth Law of Thermodynamic (d) Joule's Law

421. The compression ratio (r) is the ratio of ---------------

(a) V_1/V_2 (b) V_2/U_1 (c) $V_1 + V_2/U_1$ (d) V_1+V_2/V_2

422. For the steam, critical pressure is -------------bar.

(a) 220.2 (b) 221.2 (c) 320.2 (d) 321.2

423. The latent heat of 1 Kg steam at atmospheric pressure is--------------KJ/Kg.

(a) 2257 (b) 2157 (b) 2357 (d) 2057

424. The heating of dry saturated steam follows --------------

(a) Boyle' Law (b) Charle's Law (c) Gay Lussac Law (d) Avagadw's Law

425. Which of the following is a five tube boiler?

(a) Cochran boiler (b) Babcock and Wilcox boiler

(c) Yarrow boiler (d) Benson boiler

426. A device used to put off fire in the furnace of the boiler when the level of water in the boiler falls to an unsafe limit, is called --------------

(a) Safely Valve (b) Feed check valve (c) Blow off cock (d) Fusible plug

427. The volume of air delivered by the compressor is called--------------

(a) Free air delivery (b) Swept volume

(c) Compressor capacity (d) Compressor efficiency

428. In the process of cake making the flour is sifted two three times to------------

(a) Remove any fibre present (b) Ensure that flour is clean

(c) To incorporate air well (d) None of the above

429. Which of the following method of cooking causes maximum destruction of Vitamin A?

(a) Pressure cooking (b) Blanching (c) Frying (d) None of the above

430. When you cook food in an enclosed box as a constant tempeature, it is known as-----------

(a) Baking (b) Boiling (c) Pressure cooking (d) Steaming

ANSWER KEY

1. a	2. d	3. a	4. d	5. a	6. b	7. d	8. b	9. a	10. b
11. a	12. b	13. b	14. a	15. b	16. b	17. c	18. d	19. b	20. c
21. b	22. b	23. a	24. d	25. b	26. b	27. c	28. c	29. d	30. a
31. b	32. b	33. a	34. c	35. c	36. a	37. c	38. d	39. a	40. c
41. a	42. b	43. d	44. a	45. b	46. c	47. a	48. d	49. b	50. a
51. d	52. c	53. d	54. d	55. b	56. a	57. d	58. b	59. a	60. a
61. c	62. c	63. a	64. d	65. b	66. b	67. c	68. a	69. c	70. d
71. c	72. a	73. c	74. b	75. c	76. d	77. a	78. c	79. c	80. d
81. a	82. d	83. a	84. b	85. a	86. d	87. d	88. a	89. b	90. b
91. a	92. a	93. c	94. d	95. d	96. b	97. c	98. c	99. a	100. c
101. c	102. c	103. b	104. a	105. d	106. c	107. c	108. b	109. d	110. c
111. a	112. b	113. c	114. b	115. d	116. a	117. c	118. d	119. a	120. a
121. a	122. c	123. b	124. b	125. c	126. c	127. d	128. b	129. b	130. d
131. a	132. a	133. c	134. b	135. a	136. b	137. a	138. c	139. c	140. d
141. a	142. c	143. d	144. c	145. c	146. a	147. a	148. c	149. a	150. a
151. b	152. b	153. b	154. c	155. a	156. a	157. c	158. b	159. b	160. c
161. c	162. b	163. a	164. d	165. c	166. b	167. a	168. a	169. c	170. c
171. c	172. b	173. a	174. a	175. a	176. b	177. b	178. b	179. b	180. c
181. b	182. a	183. b	184. b	185. a	186. a	187. c	188. c	189. a	190. c
191. d	192. b	193. b	194. b	195. c	196. b	197. d	198. b	199. a	200. a
201. a	202. a	203. b	204. c	205. a	206. b	207. b	208. b	209. b	210. b
211. b	212. b	213. b	214. b	215. b	216. a	217. a	218. a	219. c	220. c
221. a	222. b	223. a	224. b	225. b	226. c	227. b	228. a	229. b	230. a
231. b	232. b	233. a	234. c	235. c	236. a	237. d	238. d	239. b	240. b
241. b	242. c	243. c	244. d	245. d	246. b	247. a	248. a	249. b	250. a
251. b	252. c	253. b	254. b	255. b	256. a	257. b	258. c	259. d	260. b
261. d	262. a	263. a	264. a	265. c	266. a	267. b	268. b	269. b	270. a
271. a	272. c	273. a	274. a	275. d	276. b	277. d	278. c	279. d	280. a
281. a	282. b	283. a	284. b	285. b	286. c	287. d	288. b	289. b	290. d
291. c	292. a	293. a	294. a	295. b	296. b	297. a	298. d	299. b	300. a
301. a	302. c	303. a	304. d	305. d	306. a	307. b	308. b	309. b	310. c
311. b	312. d	313. c	314. a	315. c	316. b	317. b	318. b	319. d	320. c
321. b	322. a	323. c	324. b	325. d	326. b	327. d	328. a	329. a	330. a
331. c	332. b	333. a	334. d	335. a	336. d	337. d	338. c	339. b	340. c
341. b	342. c	343. a	344. b	345. c	346. a	347. b	348. c	349. b	350. b
351. b	352. b	353. b	354. b	355. a	356. a	357. c	358. c	359. b	360. c
361. b	362. a	363. d	364. a	365. d	366. a	367. c	368. c	369. b	370. c

371. d	372. b	373. c	374. c	375. d	376. c	377. d	378. c	379. d	380. d
381. a	382. b	383. c	384. d	385. c	386. c	387. a	388. a	389. d	390. d
391. d	392. d	393. d	394. d	395. a	396. b	397. b	398. c	399. c	400. a
401. b	402. b	403. c	404. a	405. b	406. a	407. b	408. c	409. c	410. d
411. b	412. d	413. d	414. b	415. a	416. b	417. b	418. a	419. c	420. c
421. a	422. b	423. a	424. b	425. a	426. d	427. c	428. c	429. c	430. c

3

Food Chemistry and Nutrition

1. Ozone protects entire life from ------------- of sun.
(a) Alpha rays (b) Gamma rays (c) X rays (d) UV rays

2. ------- ---spp. are biological nitrogen fixtures present in root nodules of groundnut crop.
(a) *Rhizobium* (b) *Azotobacter* (c) *Salmonella* (d) *Lactobacillus*

3. -------------cells refer to creation of electric charge by absorption of light.
(a) Solar (b) Polar (c) Radio (d) Electromagnetic

4. ---------------- are primary cause of acid rain.
(a) Sulphur dioxide and nitrogen dioxide
(b) Ammonium oxide and ammonium carbide
(c) Hydrogen peroxide and hydrogen chloride
(d) Sodium and potassium nitrate

5. -------------- is product obtained after microbial breakdown of organic matter in the soil.
(a) Manure (b) Humus (c) Fertilizer (d) All of the above

6. ---------------- energy is utilized from interior of the earth for power generation.
(a) Geothermal (b) Biochemical (c) Bioelectrical (d) Geomagnetic

7. -------------- is the process where the amino acid gets added according to codon sequence present on mRNA.
(a) Transcription (b) Translation (c) Both a and b (d) None of the above

8. -------------- is the nitrogenous base in DNA pairs with cytosine through hydrogen bonds.
(a) Adenine (b) Guanine (c) Uracil (d) None of the above

9. -------------is the component of nucleotide other than nitrogenous base and sugar.
(a) Phosphate (b) Nitrates (c) Sulphates (d) Carbonates

10. -------- is analog of thymine.
(a) Uracil (b) Adenine (c) Cytosine (d) Alanine

11. ------------ nitrogenous bases are purines.
(a) Adenine and Guanine (b) Cytosine and thiamine
(c) Both a and b (d) None of the above

12. --------enzyme during DNA replication seals the gap after formation of new DNA fragment.
(a) DNA ligase (b) DNA lyase (c) DNA isomerase (d) DNA replicase

13. ---------- are present on DNA which transmits from generation to generation.
(a) Chromosomes (b) Genes (c) Amino acids (d) Codons

14. ------------ are the proteins having catalytic activity.
(a) Enzymes (b) Globulins (c) Albumins (d) None of the above

15. Polymer of galacturonic acid functioning as cementing material between the cells of plant tissues is called as---------------
(a) Hyluronic acid (b) Pepotides (c) Pectins (d) Cellulose

16. Diastase is made up of -------------, working on starch simultaneously.
(a) α amylase + β amylase (b) α galacturonase + β galacturonase
(c) Pectinase A and B (d) Amylose and amylopectin

17. Pernicious anemia is associated with deficiency of -------------
(a) Vit. B_1 b. Vit. B_6 c. Vit. B_{12} d. Vit. B_2

18. Non protein part of enzymes is called as-------------
(a) Holoenzyme (b) Prosthetic group (c) Rho factor (d) Isozyme

19. The organic compounds required in small amounts to perform specific biological functions for normal maintenance of organism is--------------
(a) Phospholipids (b) Glycolipids (c) Sphingolipids (d) Cholesterol

20. ------------ is called as power house of cell.
(a) Mitochondria (b) Golgi bodies (c) ER (d) Vacuoles

21. ----------- transfer amino acids to mRNA for protein biosynthesis.
(a) r-RNA (b) t- RNA (c) c-DNA (d) None of the above

22. Formation of soap is called as-------------
(a) Saponification (b) Hydrogenation (c) Glyceration (d) Esetrification

23. Formation of *ghee* is called as-------------
(a) Saponification (b) Hydrogenation (c) Glyceration (d) Esterification

24. Dry heating of sugar above their melting points in absence of amino acid to develop brown products is known as-------------
(a) Pasteurization (b) Tenderization (c) Caramelization (d) Enolization

25. Milligram of KOH required saponifying 1g of oil or fat is called as------------number.
(a) Iodine (b) Saponification (c) RM (d) Kirschner

26. Substances added to fats or fat containing foods to retard the oxidative breakdown of fatty acids during storage are known as-------------
(a) Antifoaming agents (b) Anti aging agents
(c) Antioxidants (d) Anti hazing agents

27. Development of undesirable odour in oils and fats due to lipolysis or oxidative breakdown of fatty acids is known as-------------
(a) Rancidity (b) Hydrogenation (c) Oxidation (d) None of the above

28. -------------- sugars cause gas formation in stomach.
(a) Arabinoses (b) Alohexoses (c) Raffinoses (d) Ketopentoses

29. ---------- are hair proteins.
(a) Albumins (b) Globulins (c) Lactins (d) Keratin

30. Milk is --------- type of food dispersion.
(a) Semisolid (b) Aerosol (c) Emulsion (d) Colloidal

31. ------------ is a major milk protein.
(a) Casein (b) Albumin (c) Globulin (d) All of the above

32. -------------- is a major muscle protein.
(a) Actin (b) Myosine (c) Globulin (d) Pectin

33. Gases and propellants are used as additives to produce ------------- in liquid food.
(a) Gel (b) Sol (c) Emulsion (d) Foam

34. ---------is most common organic acid used as food preservative.
(a) Benzoic acid (b) Tartaric acid (c) Citric acid (d) Hydrochloric acid

35. ------------ method is used for separation of protein.
(a) SDS PAGE (b) HPLC
(c) GLC (d) Agarose gel electrophoresis

36. ---------- is used for separation of DNA.
(a) SDS PAGE (b) HPLC
(c) GLC (d) Agarose Gel electrophoresis

37. Enzymes are purified by-----------
(a) Centrifugation (b) GLC (c) Ultra purification (d) Electrophoresis

38. --------- is called as heart of HPLC.
(a) Oven (b) Column (c) Detector (d) Solvents

39. Functioning of UV-vis spectrophotometer is based on principle of-----------law.
(a) Einstein (b) Newton (c) Edison (d) Lambert Beer

40. ----------- has highest calorific value.
(a) Carbohydrates (b) Proteins (c) Fats (d) Vitamins

41. Deficiency of ------------- cause's osetomalacia.
(a) Fe (b) Ca (c) Mg (d) Zn

42. --------- is present in chlorophyll.
(a) Fe (b) Ca (c) Mg (d) Zn

43. Which of the following is not an amino acid?
(a) Glutamic acid (b) Aspartic acid (c) Cerebronic acid (d) Serine

44. Invertase is the enzyme that hydrolyzes sucrose to------------

(a) Maltose and fructose (b) Glucose and galactose
(c) Glucose and fructose (d) Lactose and galactose

45. Which of the following is formed when glucose is oxidized to gluconic acid by the action of enzyme glucose oxidase?

(a) Hydrogen (b) Hydrogen peroxide (c) Carbon dioxide (d) Hydrogen sulfide

46. What is the minimum number of carbon atoms present in a fatty acid?

(a) Four (b) Five (c) Six (d) Seven

47. How many molecules of NADH and ATP are produced in glycolysis?

(a) 2 NADH and 2 ATP (b) 4 NADH and 4 ATP
(c) 2 NADH and 4 ATP (d) 4 NADH and 4 ATP

48. In acid modified starches, the mixing temperature of starch and water ranges between---------

(a) 28 to 32°C (b) 50 to 55°C (c) 45 to 50°C (d) 65 to 70°C

49. The circuit of conversion of lactic acid-glucose-glycogen is called as------------

(a) EMP pathway (b) Cori cycle (c) ETC pathway (d) Glyconeogenesis

50. The disease hyponatremia is caused due to --------------

(a) Higher potassium level (b) Higher sodium level
(c) Lower potassium level (d) Lower sodium level

51. Which of the following is the disorder related to blood constituents?

(a) Alkaptanuria (b) Farbers disease (c) Thalassemia (d) Galactosemia

52. Molecular formula of chitin is------------

(a) $C_8H_{11}O_5N$ (b) $C_8H_{12}O_5N$ (c) $C_8H_{13}O_5N$ (d) $C_8H_{14}O_5N$

53. The process that involves in the rearranging of acyl group in tri-acyl glycerol is called as------

(a) Hysteresis (b) β-oxidation (c) Retrogradation (d) Inter-esterification

54. Which is the naturally occurring antioxidant in foods?

(a) BHA (b) BHT (c) Tocopherol (d) PG

55. Which sugar is abundant in wheat, cotton and rice?

(a) Xylose (b) Arabinose (c) Ribose (d) Allose

56. The toxin hemagglutinin causes-------------

(a) Agglutination of proteins (b) Hemolysis of erythrocytes
(c) Agglutination of erythrocytes (d) Hemolysis of proteins

57. Tannins bind with--------------

(a) Carbohydrates (b) Lipids (c) Vitamins (d) Minerals

58. SO_2 inhibits------------

(a) Enzymatic browning (b) Non enzymatic browning
(c) Gellation (d) Crystallization

59. ------------- is produced by selective chlorination of sucrose.
(a) Saccharine (b) Sucralose (c) Aspartame (d) Alitame
60. The colour of bixin is--------------
(a) Yellow (b) Orange (c) Red (d) Green
61. The synthetic ion-exchanger in ion exchange chromatography is------------
(a) Cross linked electrolyte (b) Parallel linked electrolyte
(c) Cross linked polyelectrolyte (d) Parallel linked polyelectrolyte
62. GLC is mostly used in the analysis of ---------------compounds.
(a) Polar (b) Non polar (c) Volatile (d) Non volatile
63. Polar hydrophilic substances such as amino acid, carbohydrate and water soluble plant pigments are separated by---------------
(a) Size exclusion chromatography (b) Normal phase chromatography
(c) Reverse phase chromatography (d) Double phase chromatography
64. Ion exchange chromatography is not used for------------
(a) Ionic from non-ionic (b) Cationic from anionic
(c) Difference in partition coefficient (d) Difference in their charge
65. Paper electrophoresis was first used for protein separation by---------------
(a) Von klobusitzki and Konig (b) Ornsten
(c) Davis (d) Embden and Meyerhop
66. The ratio of intensity of transmitted light to the intensity of incident light is known as---------
(a) Isoelectric focusing (b) Transmittance
(c) Molar extinction coefficient (d) Specific extinction coefficient
67. Sulpher volatiles are responsible for odour of--------------
(a) Ginger (b) Black pepper (c) Coffee beans (d) Onion
68. The aroma of banana is due to----------
(a) 2-methyl butyric acid (b) Iso-pentyl acetate
(c) Pentyl valerate (d) Calcium propionate
69. The chemical properties of atoms are governed by its number of --------------
(a) Proton (b) Neutron (c) Electron (d) None of the above
70. Which disease is caused due to consumption of methyl-mercury contaminated fish?
(a) Itai-Itai (b) Methenoglobinemia (c) Minamata (d) Bioaccumulation
71. The opacity of emulsions is determined by-------------
(a) Droplet size
(b) Droplet cone
(c) Difference in refractive index of two phases
(d) All of the above

72. Anthocyanin is greatly affected by----------------
(a) Volume (b) pH (c) Temperature (d) Volume and pH
73. Flavonoides are having --------------- absorption characteristics.
(a) Light (b) Water (c) Wave (d) Colour
74. Proanthocyanides are----------------
(a) Yellow in colour (b) Colourless (c) Blue in colour (d) Black in colour
75. Vitamins are required for various ----------------- reactions.
(a) Non-metabolic (b) Metabolic (c) Irreversible (d) Reversible
76. Which enzyme is used for tenderization of meat?
(a) Papain (b) Protease (c) Amylase (d) Lipoxygenase
77. ------------- pollution affects photosynthesis and destroys chlorophyll by entering through stomata.
(a) Water (b) Sound (c) Air (d) Soil
78. Pollution is a --------------- change in the physical, chemical and biological characteristics.
(a) Undesirable (b) Desirable (c) Chemical (d) Gaseous
79. Essential fatty acid serves as precursors of ---------------
(a) Vitamin C (b) Prostaglandin (c) Niacin (d) Retinol
80. Ovomucoid is the protein found in----------
(a) Fruit (b) Vegetables (c) Milk (d) Egg
81. Colourant used in butter is------------------
(a) Annato (b) Erythrocyte (c) Congo red (d) None of the above
82. The nucleic acid of a virus is---------------
(a) DNA only
(b) RNA only
(c) Both DNA and RNA
(d) Either DNA or RNA
83. The smallest unit of heredity is -------------
(a) Chromosome (b) Gene (c) Codon (d) Nucleotide
84. Chromanol is basic ring structure of ------------ vitamin.
(a) A (b) C (c) D (d) E
85. Ninhydrin test is used for -------------
(a) Protein (b) Carbohydrate (c) Fats (d) Minerals
86. Digestion of carbohydrate starts from-----------
(a) Mouth (b) Stomach (c) Duodenum (d) Small intestine
87. In non- competitive inhibition Km------------
(a) Do not change
(b) Decreases
(c) Increases
(d) Increases or decreases depending upon the type of enzymes

88. Which of the following is plant proteinase?
(a) Pepsin (b) Trypsin (c) Urease (d) Papain
89. Kjeldahl's method is used for estimation of--------------
(a) Crude fiber (b) Crude fat (c) Crude protein (d) None of the above
90. The reagent used in the Selivanoff's test is------------
(a) Resorcinol (b) Napthol (c) Kojic acid (d) Potassium iodate
91. Cholesterol is synthesized from-------------
(a) Alanine (b) Stearic acid (c) Acetyl coenzyme (d) Choline
92. Egg white is rich in------------
(a) Carbohydrate (b) Minerals (c) Protein (d) Fat
93. DNA and RNA are long linear polymers called --------------
(a) Nucleotides (b) Nucleic acids (c) Genes (d) Nitrogen bases
94. Back bone of the nucleic acids is known as-------------
(a) Ribose sugar (b) Nitrogen base (c) Glycosidic bond (d) Phosphodiester linkage
95. DNA and its associated protein is called-------------
(a) Chromosome (b) Gene (c) Nucleic acid (d) All of the above
96. During replication, two strands of DNA are separated by enzymes termed as------------
(a) DNA ligase (b) DNA polymerase (c) Helicase (d) Nuclease
97. DNA synthesis is catalyzed by-------------- enzyme.
(a) DNA polymerases (b) RNA polymerases
(c) DNA-RNA polymerases (d) Nucleases
98. Synthesis of RNA from DNA template is known as--------------
(a) Translation (b) Transcription (c) Protein synthesis (d) Transformation
99. The amino acids attached to the tRNA by enzyme------------
(a) Aminoacyl t-RNA synthetases (b) Polymerases
(c) Endonucleases (d) Peptidyltransferas
100. Proteins are synthesized by linking amino acids with -------------- bonds.
(a) Disulfide (b) Glycosidic (c) Peptide (d) Ester
101. The base pairs are held together primarily by------------------
(a) Covalent bonds (b) Hydrogen bonds (c) Ionic bonds (d) Vander walls forces
102. ------------- is not a saturated fatty acid.
(a) Palmitoleic (b) Palmitic (c) Propionic (d) Arachidic
103. --------------- is the largest macromolecules in the cell.
(a) Nucleic acid (b) Proteins (c) Lipids (d) Carbohydrates
104. --------------- are the primary products of photosynthesis.
(a) Carbohydrates (b) Lipids (c) Proteins (d) Vitamins

105. --------------- is structural and functional unit of life.
(a) Tissue (b) Cell (c) Organ (d) Organ system

106. The basic unit of photosynthetic apparatus is called as-------------
(a) Thylakoid (b) Mesosomes (c) Cytoplasm (d) Glycocalyze

107. The external non protoplasmic covering of cell is---------------
(a) Cell membrane (b) Plasma membrane (c) Cell wall (d) Capsule

108. ------------- is aqueous portion of the cytoplasm.
(a) Fluid (b) Ectoplasm (c) Cytosol (d) Watery fluid

109. Endoplasmic reticulum is discovered by-----------------
(a) Camilo Golgi (b) Keith Porter (c) de Duve (d) Robert Hooke

110. ------------- is a marker enzyme for golgi apparatus.
(a) Cell membrane (b) Mitochandria (c) Lysosome (d) Golgi body

111. The vacuoles of plant cell are covered by single membrane called-------------
(a) Tonoplast (b) Contractile (c) Cell sap (d) Tinoplast

112. ---------------- are involved in purine metabolism.
(a) Glyoxysome (b) Peroxisomes (c) Lysosome (d) Mesosomes

113. Number of ATP produced in EMP pathway are---------------
(a) 32 (b) 08 (c) 24 (d) 12

114. α – amylase enzyme is found in the-------------- of animals.
(a) Blood (b) Stomach (c) Saliva (d) Kidney

115. Coenzyme Q_{10} is known as ------------
(a) Quinone (b) Flavoprotein (c) Ubiquinone (d) Isoprenon

116. TPP stands for --------------
(a) Treptone polyphosphate (b) Thiamine pyrophosphate
(c) Thiamine polyphosphate (d) Tyrosin pyrophosphate

117. Citric acid cycle was discovered by ----------------
(a) Hans Krebs (b) Embed (c) Robert Koach (d) Fisher

118. Reducing sugars are identifying by which test?
(a) Biuret (b) Benedict's (c) Ninhydrin (d) Inversion

119. Mucopolysaccharides are called as --------------
(a) Glycosaminoglycans (b) Mucoids
(c) Proteoglycans (d) Heteroglycan

120. The functions of ligase is to -----------
(a) Rejoin segments of DNA (b) Make longitudinal cuts in DNA
(c) Synthesize cDNA (d) Break down ligaments

121. rRNA is the major component of ------------
(a) Mitochondria (b) Ribosomes (c) Chloroplast (d) Nucleus

122. All enzymes are proteins, but all proteins are not------------
(a) Enzymes (b) Proteins (c) Precursors (d) Biocatalyst

123. ------------ is father of biochemistry.
(a) Robert Roach (b) Carl Neuberg (c) Antonine Lavoiser (d) Nicohl Appert

124. ------------is the defense protein.
(a) Immunoglobulin (b) Glutelin (c) Insulin (d) Serum

125. ------------- is the complete proteins.
(a) Egg albumin (b) Lecithin (c) Cephalin (d) Legumin

126. ----------- is an antirachitic vitamin.
(a) Vitamin D (b) Vitamin A (c) Vitamin C (d) Vitamin B_1

127. Xerophthalmia is caused due to deficiency of -------------vitamin.
(a) D (b) C (c) A (d) E

128. ------------ is antipernicious anemia vitamin.
(a) Vitamin D (b) Vitamin A (c) Vitamin K (d) Vitamin B_{12}

129. ---------- is blood clotting factor.
(a) Prothrombin (b) Thrombin (c) Cobalmin (d) Quionone

130. Enzyme involved in catalyzes transfer of functional group is-------------
(a) Oxidoreductase (b) Transferase (c) Hydrolasea (d) Lyases

131. -------------- is the nucleotide coenzyme.
(a) NAD (b) ATP (c) CDP (d) TPD

132. Heparin is -------------- type carbohydrate.
(a) Heteropolysaccharide (b) Diasaccharide
(c) Trisaccharide (d) Tetrasaccharide

133. The location of TCA cycle is------------
(a) Cytasol (b) Liver (c) Stomatch (d) Mitochondrial matrix

134. The synthesis of glucose or glycogen from carbohydrate compounds is known as----------
(a) Gluconeogenesis (b) Glycolysis
(c) TCA (d) HMP

135. ------------- in foods is to serve as fuel molecules.
(a) Lipids (b) Proteins (c) Vitamins (d) Carbohydrate

136. ------------- is the phenomenon in which a substance occurs in different crystalline forms.
(a) Polymer (b) Polymorphism (c) Tempering (d) Hydrogenation

137. -------------- is the essential fatty acid.
(a) Oleic acid (b) Stearic acid (c) Linoleic acid (d) Palmitoleic

138. --------------- occur as constituents of plant cell wall and middle lamella.
(a) Pectic substances (b) Hemicellulose (c) Cellulose (d) Gums

139. The product of partial breakdown of starches is-----------
(a) Glycogen (b) Cellulsoe (c) Dextrin (d) Gum

140. -------------- consists of glucose units linked by α1-4 and α1-6 bonds.
(a) Amylose (b) Amylospectin (c) Hemicellulose (d) Glucogen

141. ------------ are the acyl derivatives of glycerol.
(a) Triglycerides (b) Diglyceride (c) Monoglyoride (d) None of the above

142. ------------- are the unsaturated fatty acids.
(a) Myristoleic and Arachidonic (b) Palmitoleic and Linolenic
(c) Oleic and Linoleic (d) All of the above

143. Following is not saturated fatty acid.
(a) Bytyric (b) Oleic (c) Capric (d) Lauric

144. Molarity is expressed in-------------
(a) Gm/lit (b) Mol/lit (c) N/lit (d) Kg/lit

145. A specific three dimensional configuration involved in ------------- structure of protein.
(a) Primary (b) Tertiary (c) Secondary (d) Quaternary

146. Km is -------------
(a) Equilibrium constant (b) Michaelis constant
(c) Mortar constant (d) None of the above

147. -------------- are the condensation products of sugar with non sugars.
(a) Peptide bond (b) Ester linkage (c) Glycoside (d) Glycogen bond

148. -------------- a phospholipid is very good emulsifier.
(a) Lecithin (b) Glycerol (c) Cephalin (d) Acytol

149. Environmental factors such as temperature and pH exert their effect on the -----------of microbial cells.
(a) Membranes (b) DNA (c) Enzymes (d) Cell wall

150. An enzyme ------------ the activation energy required for a chemical reaction.
(a) Increases (b) Provides (c) Lowers (d) Catalyzes

151. Algin is a-----------
(a) Polysaccharide (b) Lipid (c) Protein (d) Provitamin

152. An apoenzyme is where ------------- is located.
(a) Cofactor (b) Coenzyme (c) Redox reaction (d) Active site

153. Many coenzymes are------------
(a) Metals (b) Vitamins (c) Proteins (d) Substrates

154. To digest cellulose in its environment, a fungus produces a/an---------
(a) Endoenzyme (b) Exoenzyme (c) Catalase (d) Polymerase

155. In negative feedback control of enzymes, a buildup in the amount of ---------- decreases the activity of the enzyme.
(a) Substrate (b) Reactant (c) Product (d) ATP

156. Two sugars differing only in configuration around one specific carbon atom are called as----
(a) Anomers (b) Epimers (c) Isomers (d) Conformers

157. Isomeric forms of monosaccharide that differs only in the configuration about carbonyl alcohol are called as-----------
(a) Anomers (b) Epimers (c) Isomers (d) Conformers

158. In which of the following lipo-proteins, the major portion is cholesterol?
(a) VLDL (b) LDL (c) HDL (d) All of the above

159. Which of the following sugar(s) is/are present in honey?
(a) Levulose (b) Fructose (c) Both a and b (d) None of the above

160. Which of the following is liquefying enzyme?
(a) α- amylase (b) β- amylase (c) Both a and b (d) None of the above

161. Lactose is -------------
(a) Monosaccharide (b) Disaccharide (c) Oligosaccharide (d) Polysaccharide

162. Which of the following is the saccharifying enzyme?
(a) α- amylase (b) β- amylase (c) Both a and b (d) None of the above

163. Which of the following compound/s do you expect at the end of Strecker degradation?
(a) Acid (b) Ester (c) Aldehyde (d) All of the above

164. Which of the following is/are storage polysaccharide?
(a) Dextran (b) Fructan (c) Starch (d) All of the above

165. Which of the following is/are the structural polysaccharide?
(a) Chitin (b) Cellulose (c) Lignin (d) All of the above

166. Which of the following is the polymer of aromatic alcohol?
(a) Lignin (b) Hemicelluloses (c) Pectin (d) None of the above

167. Disaccharide produced after hydrolysis of cellulose is/are------------
(a) Glucose (b) Maltose (c) Cellubiose (d) All of the above

168. Which of the following sugars is/are non-reducing?
(a) Sucrose (b) Cellubiose (c) Maltose (d) All of the above

169. The most common carbohydrate monomer is-----------
(a) Maltose (b) Lactose (c) Glucose (d) Galactose

170. Simplest form of carbohydrate is--------------
(a) Carbon (b) Starch (c) Monosaccharide (d) Cane sugar

171. Which of the following is/are responsible for the delay in gelatinization?
(a) Fat (b) Protein (c) Sugar (d) All of the above

172. Which lipid is saponifiable?
(a) Simple (b) Complex (c) Both a and b (d) None of the above

173. Which lipid does not contain fatty acids?
(a) Simple (b) Complex (c) Both a and b (d) None of the above

174. Phosphoglycerides are the examples of-----------
(a) Simple lipid (b) Complex lipid (c) Carbohydrates (d) Steroids

175. Steroids come under -------------- categories of lipids.
(a) Simple (b) Complex (c) Compound (d) None of the above

176. Waxes come under ---------------categories of lipids.
(a) Simple (b) Complex (c) Both a and b (d) None of the above

177. Which of the following is the essential fatty acid?
(a) Stearic acid (b) Linolenic acid (b) Lauric acid (d) Palmitic acid

178. Arachidonic acid contains --------------- carbon atoms.
(a) 16 (b) b. 18 (c) 20 (d) 22

179. Triglyceride, which are solid at room temperature are often referred as------------
(a) Fat (b) Oil (c) Steroid (d) Waxes

180. Triglyceride, which are liquid at room temperature are often referred as------------
(a) Fat (b) Oil (c) Steroid (d) Waxes

181. Ovomucoid is the protein found in--------------
(a) Fruits (b) Vegetables (c) Milk (d) Eggs

182. Which vitamin is absent in egg?
(a) Vit. A (b) Vit. C (c) Vit. D (d) Vit. K

183. For hydrogenation of oils the catalyst required is-------------
(a) Fe (b) Al (c) Ni (d) Mg

184. Most abundant mineral present in the egg shell is-----------------
(a) Fe (b) Mg (c) Zn (d) Ca

185. Starch gel is-------------
(a) Pseudoplastic (b) Plastic (c) Elastic (d) Thixotropic

186. Which fatty acid helps in improvement of vision?
(a) DHA (b) EPA (c) Linoleic acid (d) Stearic acid

187. Milk contains which sugar?
(a) Sucrose (b) Starch (c) Cellulose (d) Lactose

188. Lactose increases the retention of-------------
(a) Ca (b) P (c) Fe (d) I

189. Which vitamin is known to have antisterility factor in rats?
(a) A (b) D (c) E (d) K

190. Vitamin D helps in absorption of-------------

(a) Ca (b) P (c) Both a and b (d) None of the above

191. Which of the following vitamin is water soluble?

(a) A (b) C (c) D (d) E

192. Which of the following vitamin is responsible for clotting of blood?

(a) B (b) K (c) C (d) A

193. Acid-alkali reaction of the blood is controlled by-------------

(a) P (b) I (c) K (d) Na

194. Copper aids in the utilization of ----------------- and hemoglobin synthesis.

(a) Fe (b) Mg (c) P (d) Mb

195. Which vitamin is known as "sunshine vitamin?"

(a) A (b) C (c) D (d) E

196. Which of the following is water soluble vitamin?

(a) Vitamin K_1 (b) Vitamin K_2 (c) Vitamin K_3 (d) All of the above

197. ------------- is another term for biosynthesis.

(a) Catabolism (b) Anabolism (c) Metabolism (d) Catalyst

198. Which of the following is a reduced compound?

(a) NAD (b) FAD (c) NADH (d) ADP

199. Products of glycolysis are--------------

(a) ATP (b) ADP (c) CO_2 (d) NADH

200. The net yield of ATP's in complete oxidation of glucose in aerobic respiration is -----------

(a) 40 (b) 6 (c) 38 (d) 2

201. Polishing of rice removes --------------vitamin.

(a) K (b) B_1 (c) C (d) A

202. Bactomatic is a method to determine the bacterial---------------

(a) Activity (b) Size (c) Reproduction (d) Decimal point value

203. Temperature at which fat rises in a cylinder, is known as------------

(a) Slip point (b) Melting point (c) Softening point (d) None of the above

204. Peroxide value is the measure of--------------quality.

(a) Protein (b) Fat (c) Sugar (d) Vitamin

205. For soybean oil, peroxide value of -------------- or less indicates freshness.

(a) 1 (b) 10 (c) 100 (d) 1000

206. Based on iodine value, which of the following oil is best for human nutrition?

(a) Palm oil (b) Groundnut oil (c) Olive oil (d) Cottonseed oil

207. The small molecules of double stranded, helical, non-chromosomal DNA are known as------

(a) Gene (b) Plasmids (c) F factor (d) Protoplast

208. Polymerase Chain Reaction is invented by-----------
(a) David Ehrenfeld (b) Kary B. Mullis (c) Mery B. Mullis (d) Peter J. Russel

209. Which of the following food component supply major share of energy to the human body?
(a) Lipids (b) Carbohydrates (c) Proteins (d) Vitamins

210. Which of the enzymatic test is used to detect inadequate pasteurization?
(a) Lipase (b) Phosphatase (c) Peroxidase (d) Catalase

211. Which acid is produced as a result of activity of insects and rodents during the storage of farm produce?
(a) Oxalic acid (b) Uric acid (c) Citric acid (d) Malic acid

212. The major antinutritional factor present in rapeseed and mustard is--------------
(a) Oxalic acid (b) Ellagic acid (c) Erucic acid (d) Behenic acid

213. The end product of purine catabolism in normal humans is-------------
(a) Uric acid (b) Urea (c) Creatinine (d) Xanthine

214. In cellulose polysaccharide, glucose units are linked together by a glycosidic bond of-------------- linkages.
(a) β-1,4 (b) α-1,3 (c) α-1,4 (d) α-1,6

215. In which structural level, amino acids are linked together by peptide bonds in linear arrangements?
(a) Primary (b) Secondary (c) Tertiary (d) Quaternary

216. An enzyme catalyzes the conversion of aldose sugar to a ketose sugar is classified as-------
(a) Oxidoreductase (b) Transferase (c) Hydrolase (d) Isomerase

217. Cholesterol can act as a precursor for each of the following compounds except---------
(a) Chenodeoxycholic acid (b) Testesterone
(c) Glycocholic acid (d) Cholecystokinin

218. Increased protein use is accompanied by an increased dietary requirement for--------vitamin.
(a) B_2 (b) C (c) B_{12} (d) B_6

219. Which nutritional disorders are widely present in India?
(a) Anemia (b) Blindness
(d) Protein–energy malnutrition (d) All of the above

220. High fructose syrup production from starches of various origins is catalyzed by--------------
(a) Amylase (b) Glucoseoxidase (c) Invertase (d) All of the above

221. Cereal proteins are generally limiting in--------------
(a) Methionine (b) Lysine (c) Tryptophan (d) All of the above

222. The preparation of protein isolates from legume/oilseed meals involve------------
(a) Acidic extraction followed by isoelectric precipitation
(b) Alkaline extraction followed by isoelectric precipitation
(c) Neutral extraction followed by isoelectric precipitation
(d) Alkaline extraction followed by salt precipitation

223. Dietary deficiency of ascorbic acid leads to the nutritional disorder called as--------------
(a) Beri-beri (b) Scurvy (c) Marasmum (d) Goitre

224. pH meter has-------------- cell(s)
(a) One (b) Two (c) Three (d) None of the above

225. The table salt fortification with iodine is done to prevent--------------
(a) Scurvy (b) Goiter (c) Anemia (d) Pellagra

226. Lactose intolerance observed in some infants is due to----------------
(a) Genetic error (b) Poor health of mother
(c) Poor health of infant (d) Defective milk of mother

227. Which of the following sugar is not digested in human system?
(a) Sucrose (b) Maltose (c) Raffinose (d) Glucose

228. Synthetic sweeteners are commonly used as sweetening agents in food because------------
(a) They are easily available
(b) The sweetness response is faster
(c) They have long shelf life
(d) They have low calorific value per unit of sweetness

229. Legume proteins are most limiting in--------------
(a) Lysine (b) Methionine (c) Phenylalanine (d) Threonine

230. Peroxide value of stored oil indicates a degree of----------------
(a) Hydrolysis (b) Reversion of oil (c) Oxidation of oil (d) Lipolysis of oil

231. The major sugar of animal origin in our food supply is--------------
(a) Sucrose (b) Lactose (c) Glucose (d) Mannose

232. The most abundant unsaturated fatty acids in cow milk fat is--------------
(a) Oleic (b) Linoleic (c) Searic (d) Pamitic

233. The most predominant fatty acid in sunflower oil is---------------
(a) Oleic (b) Linolenic (c) Linoleic (d) Arachiodonic

234. Oil considered to be safe for consumption, if its peroxide value is less than-------- meq/kg oil.
(a) 5 (b) 10 (c) 20 (d) 40

235. The toxic substance present in cotton seed is-------------
(a) Gossypol (b) Saponins (c) Goitrogens (d) Hamagglutanins

236. Deficiency of nicotinic acid leads to a disease called------------
(a) Beriberi (b) Marasmus (c) Pellagra (d) Anaemia

237. The enzyme widely used in tenderization of meat is------------
(a) Amylase (b) Lipase (c) ATPase (d) Papain

238. National Institute of Nutrition is located at-------------
(a) Mysore (b) Mumbai (c) Delhi (d) Hyderabad

239. Keratomalacia is a disease caused due to the deficiency of------------

(a) Fat (b) Vitamin A (c) Protein (d) Calcium

240. Dietary fibre includes-----------

(a) Cellulose (b) Hemicellulose (c) Pectin (d) All of the above

241. Gluten is important protein of--------------

(a) Wheat (b) Maize (c) Rice (d) Sorghum

242. The pH of 1mM solution of sodium hydroxide will be------------

(a) 14 (b) 11 (c) 8 (d) 3

243. Which of the following is not energy yielding nutrient?

(a) Vitamin (b) Protein (c) Carbohydrate (d) Fat

244. Starch is polymer of----------------

(a) Sucrose (b) Glucose (c) Fructose (d) Galactose

245. Which quality index is associated with the development of rancidity in fats and oils?

(a) Iodine number (b) Saponification number

(c) Peroxide number (d) None of the above

246. Hydrolytic enzymes are mainly present in which cell organelle?

(a) Mitochondria (b) Nucleus (c) Golgi bodies (d) Lysosomes

247. Which of the following bond is present in DNA molecule?

(a) Glycosidic (b) Phosphodiester (c) Peptide (d) Ester

248. The genetic information of an organism is stored in which organelle?

(a) Cytoplasm (b) Nucleus (c) Mitochondria (d) Golgi bodies

249. How many net ATPs will be produced when one molecule of glucose is oxidized to pyruvate via aerobic glycolysis?

(a) 2 (b) 4 (c) 6 (d) 8

250. Which biomolecules acts as a connecting link between glycolysis and citric acid cycle?

(a) Acetyl CoA (b) Glucose (c) Pyruvate (d) Glyceraldehyde

251. --------------is an example of reducing disaccharide.

(a) Sucrose (b) Maltose (c) Trehalose (d) Glucose

252. -------------- is the major component of the endosperm of rice.

(a) Protein (b) Starch (c) Fat (d) Vitamins

253. The principle protein of rice is-------------

(a) Gluten (b) Zein (c) Oryzenin (d) None of the above

254. Oils from cereal grains are rich in which vitamin?

(a) A (b) B (c) E (d) D

255. Corn gluten is rich in the corn protein known as---------------

(a) Zein (b) Gluteinin (c) Glutelin (d) Prolamin

256. -------------- is the study of foods and food constituents on gene expression.
(a) Pretromics (b) Nutrigenomics (c) Nutrition (d) None of the above

257. β-glucans are present in high amount in--------------
(a) Oat bran (b) Rice bran (c) Barley (d) Nutmeg

258. Omega-3 fatty acid, the neutraceutical is from which source?
(a) Plant (b) Animal (c) Microbial (d) None of the above

259. Pulses are the major source of ------------- to vegetarian masses of the country.
(a) Protein (b) Carbohydrate (c) Fat (d) Minerals

260. Disaccharide produced after hydrolysis of celluloses is / are-----------
(a) Glucose (b) Maltose (c) Cellubiose (d) All of the above

261. Which of the following is the biotin binder?
(a) Avidin (b) Aflatoxin (c) Gossypol (d) Ovalbumin

262. Deficiency of which vitamin leads to rickets.
(a) Vit. D (b) Vit. A (c) Vit. E (d) Vit. C

263. The normal salt intake to avoid complications like hypertension, should be ------------g/day.
(a) 2.5 (b) 7.8 (c) 5.0 (d) 1.2

264. ----------- vitamin is absent in lean meat.
(a) Vitamin A (b) Vitamin B (c) Vitamin C (d) Vitamin D

265. The enzymatic activity of egg yolk is associated with------------
(a) Phosvitin (b) Livetin (c) Lipovitellin (d) None of the above

266. The native enzyme in meat which helps in tenderization during cold storage is-----------
(a) Lipase (b) Myosinase (c) Collagenase (d) Cathepsins

267. The method of enzyme immobilization that has high loading per unit of matrix is----------
(a) Adsorption (b) Covalent binding (c) Entrapment (d) Membrane confinement

268. Which amylase is usually used in bread making?
(a) Bacterial (b) Cereal (c) Fungal (d) None of the above

269. The most commonly used enzyme for clarification of fruit juices is--------------
(a) Pectinase (b) Lactase (c) Protease (d) Catalase

270. The most commonly used immobilized enzyme for removal of bitterness in citrus juice is---
(a) Lipase (b) Naringinase (c) Glucose isomerase (d) Amylases

271. Which enzyme is involved in browning of cut surfaces of fruits or vegetables?
(a) Peroxidase (b) Polyphenol oxidase (c) Catalase (d) Amino acid oxidase

272. An enzyme that catalyzes the hydrolysis of pectins into pectic acids and methanol is-------
(a) Amylase (b) Cellulase (c) Pectinesterase (d) None of the above

273. An enzyme that catalyzes the hydrolysis of glucosidic linkage of polymerized galacturonic acids is---------------
(a) Polyphenol oxidase (b) Peroxidase
(c) Pectin esterase (d) Polygalacturonase

274. Argemone oil content which toxic substance?
(a) Solanine (b) Sanguinarine (c) BOAA (d) Pyrimidine bases

275. Transcription is the process of--------------
(a) Formation of RNA to DNA (b) Formation of DNA to RNA
(c) Both a and b (d) None of the above

276. Oil that contain linolenic acid are subject to flavour----------------
(a) Enhancement (b) Reversion (c) Suppression (d) Putrefaction

277. Ribosomes are the centres of ------------synthesis.
(a) Protein (b) Amino acid (c) Nucleic acid (d) Enzyme

278. ---------------is antinutritional factors in legumes.
(a) Goiter (b) Vit B complex (b) Protease inhibitor (d) None of the above

279. ----------- causes hypothyroidism.
(a) Goitrogens (b) Cyanins (c) Kunitz inhibitor (c) Both b and c

280. Lathyrogens causes-------------
(a) Goitrogens (b) Neurolathyrism (c) Cancer (d) Both a and b

281. Saponins have ability to form---------------
(a) Gel (b) Crystals (c) Foam (d) None of the above

282. Cyanogens is found in some---------------
(a) Animal food (b) Plant food (c) Flowers (d) Stem

283. ----------- is the anti nutritional factor agglutinate blood cells.
(a) Hemagglutinn (b) Cynogens (c) Saponins (d) Goitrogen

284. -------------- is a nervous disease that cripples man.
(a) Lathyrism (b) Hemagglutins (c) Inlargement (d) Urine store

285. A neurotoxin present in lathyrus *dhal* is---------------
(a) Phenolic glucoside (b) Glucosinolates
(c) Cynogenic glucoside (d) BOAA

286. Green chilli is adulterated with-----------
(a) Metanil yellow (b) Malachite green (c) Chlorophyll (d) None of the above

287. When muriatic acid is added to jiggery effervescences created indicates the presence of ----
(a) Na- bicarbonate (b) Na- carbonate (c) Na- hydroxide (d) Both a and b

288. Addition of 2 drops of bromothymol blue to milk results in development of blue colour indicates the presence of ---------------- in milk.
(a) Urea (b) Na- carbonate (c) Na- hydroxide (d) None of the above

289. Dry red chilli is adulterated with-------------
(a) Rhodamine B colour (b) Indigo carmine
(c) Both a and b (b) None of the above

290. Enzyme kinetics is study of-------------

(a) Enzyme immobilization (b) Enzyme activation

(c) Enzyme inhibition (d) Rate of reaction catalyzed by enzyme

291. -------------- is used for measuring calorific value of solid fuels.

(a) Barrel calorimeter (b) Bomb calorimeter

(c) Junkers calorimeter (d) All of the above

292. --------- is the oldest separation process and most widely used unit separation in industry on the basis of volatilities of component.

(a) Filtration (b) Centrifugation (c) Distillation (d) All of the above

293. The kinetics of most enzyme reactions are reasonably well represented by---------------

(a) Newton's Equation (b) Michaelis Menten Equation

(c) Arrhenius Equation (d) None of the above

294. The stability of enzymes is frequently reported in terms of--------------

(a) Half life (b) Full life (c) Shelf life (d) All of the above

295. Solid – liquid extraction is an example of a ------------ operation.

(a) Evaporation (b) Separation (c) Absorption (d) None of the above

296. Rate of enzyme deactivation is strongly dependent on-------------

(a) Temperature (b) Humidity (c) Thermal death time (d) All of the above

297. Enzymatic decomposition of protein, peptide or amino acid may result in the abnoxious odour termed as--------------

(a) Putrefaction (b) Rancidity (c) Decay (d) Deproteination

298. Which of the following is the polymer of aromatic alcohol?

(a) Lignin (b) Hemicelluloses (c) Pectin (d) None of the above

299. Anabolic process of converting extra glucose into glycogen is called as-----------

(a) Catabolism (b) Metabolism (b) Anabolism (d) Glycogenesis

300. Sweet curdling in milk is due to which enzyme?

(a) Rennet (b) Caseinase (c) Proteinase (d) Lipases

301. ---------- are the oligonucleotides used to detect the complementary nucleotide sequence present on genome.

(a) Probe (b) RNA (c) Nucleotides (d) Primers

302. Who proposed the structural model of DNA?

(a) Watson and Crick (b) Jenner

(c) Flemming (d) Waksman

303. Transfer of DNA on nitrocellulose membrane is referred as--------------blotting.

(a) Western (b) Southern (c) Northern (d) Dot

304. --------------- is the toxic heavy metal present in food chain.

(a) Hg (b) Fe (c) Arsenic (d) None of the above

305. The compounds reduces the surface tension and helps in removal of soil particles in cleaning compound are known as----------------

(a) Sequestrants (b) Surfactants (c) Biocides (d) None of the above

306. Which amino acid is used as flavour enhancer in food industry?

(a) Citric acid (b) Glutamic acid (c) Aspartic acid (d) All of the above

307. Invert sugar is equimolar mixture of glucose and------------

(a) Maltose (b) Sucrose (c) Galactose (d) Fructose

308. Buffers when used in fermentation media functions as----------------

(a) Changes the pH (b) Resist the change in pH

(c) Changes pH towards acidic (d) Makes pH neutral

309. Rhubarb leaf poisoning is due to ---------------- content of leaves.

(a) Oxalic acid (b) Acetic acid (c) Citric acid (d) Malic acid

310. Which of the following enzyme is used for the random digestion of pectin?

(a) Pectin esterase (b) Pectin lyase (c) Pectin esterase (d) Polygalacturonase

311. Cyanocobalamin is the name of which vitamin?

(a) Vit. B_3 (b) Vit.B_5 (c) Vit.B_{12} (d) Vit.B_7

312. Transfer of genes from one cell to another by a bacteriophase is known as------------

(a) Conjugation (b) Transduction (c) Transformation (d) None of the above

313. The organism which uses organic compounds as electron donor is known as-------------

(a) Organotrophs (b) Liphotrophs (c) Chemolithotrophs (d) None of the above

314. In man, DNA is---------------

(a) Double stranded (b) Three stranded (c) Four stranded (d) Single stranded

315. The cell without cell wall is called as--------------

(a) Cytoplast (b) Organoplast (c) Protoplast (d) None of the above

316. -------------catalyzes transcription of RNA nucleotide sequence into complementary DNA.

(a) Reverse transcriptase (b) Restriction endonuclease

(c) Ligases (d) Nucleases

317. Scouting compounds are also known as--------------

(a) Chemical abrasives (b) Physical abrasives

(c) Surface active agents (d) Cleaning compounds

318. ---------------is produces inside and secreted outside the cell.

(a) Endoezyne (b) Exoenzyme (c) Both a and b (d) None of the above

319. ------------is used as a precursor in the production of antibiotics.

(a) Phenyl acetic acid (b) Phenol

(c) Formalin (d) None of the above

320. ---------------- enzyme is used for the inversion of sugar.
(a) Invertase (b) Amylase (c) Protease (d) None of the above

321. Which of the following in not an antibiotic?
(a) Penicillin (b) Streptomycin (c) Tetracylin (d) Quinine

322. Which one of the following is an example of primary metabolites?
(a) Antibiotics (b) Amino acids (c) Enzyme inhibitors (d) Pigments

323. -------------- is the shrinkage of cell contents as a result of withdrawal of water by osmosis.
(a) Plasmolysis (b) Osmolysis (c) Autolysis (d) All of the above

324. Extra chromosomal DNA which exist in covalently closed circular form is------------
(a) Plasmid (b) Prophage (c) Both a and b (d) None of the above

325. Hypochlorites are more effective at----------------
(a) Acidic pH (b) Neutral pH (c) Alkaline pH (d) None of the above

326. The non-volatile compound in oak wood barrel is--------------
(a) Phenolic acid (b) Phenol aldehyde (c) Vanillin (d) Eugenol

327. The end product of vinegar oxidative fermentation is------------
(a) Lactic acid (b) Pyruvic acid (c) Ethanol (d) Acetic acid

328. In preparation of curd from milk the fermentation is terminated due to------------
(a) Complete oxidation of milk sugar (b) Death of fermenting organisms
(c) Fall in pH which inhibit microbial activity (d) None of the above

329. Which of the following metal is not highly toxic?
(a) Sodium (b) Lead (c) Arsenic (d) Cadmium

330. The symptoms of cadmium toxicity involves-------------
(a) Excessive salivation (b) Prostate cancer
(c) Chills (d) Both a and b

331. Ulceration in oral cavity is due to deficiency of------------
(a) Vitamin A (b) Vitamin C (c) Thiamine (B_1) (d) Pyridoxine (B_6)

332. Of the 150 pounds of food additives we eat every year, 147 pounds are-------------
(a) Salts and sweeteners (b) Artificial colourings
(c) Emulsifiers (d) Natural flavourings

333. Which of the following does not contain protein?
(a) Pili (b) Flagellum (c) Lipoteichoic acid (d) Porin

334. A diode array detector (DAD) coupled with UV detection is advantageous because-----------
(a) It allows a single wavelength of detection to be more precisely chosen
(b) It covers a range of wavelengths
(c) It allows lower concentrations of analyte to be detected
(d) It speeds up the detection at a single wavelength

335. In reversed-phase HPLC--------------
(a) A hydrophilic stationary phase is combined with a polar mobile phase
(b) A hydrophilic stationary phase is combined with a non-polar mobile phase
(c) A hydrophobic stationary phase is combined with a polar mobile phase
(d) A hydrophobic stationary phase is combined with a non-polar mobile phase

336. Lead levels in drinking water could be determined by using--------------
(a) HPLC (b) AAS (c) GC-MS (d) CZE

337. Which of the following techniques would be useful for monitoring dioxin levels in the environment?
(a) GC-MS (b) AES (c) AAS (d) TLC

338. Which statement is incorrect about the SDS-PAGE technique?
(a) Separations are run on gel-covered plates
(b) Sodium dodecyl sulfate is used to denature the samples being separated
(c) It is used to separate proteins
(d) It is a type of gas chromatographic method of separation

339. The Beer-Lambert Law gives a linear correlation with positive gradient between-----------
(a) Wavelength and absorbance
(b) Absorbance and concentration
(c) Molar extinction coefficient and absorbance
(d) Molar extinction coefficient and concentration

340. Which of the following is commonly used as the mobile phase in GC-MS?
(a) He (b) Air (c) CH_3CN (d) H_2O

341. What is the density of an object with a volume of 15 mL and a mass of 42g?
(a) 0.352 g/mL (b) 2.80 g/mL (c) 630 g/mL (d) None of the above

342. Mass spectrometers separate isotopes of different elements based on their----------
(a) Mass (b) Electric charge
(c) Mass divided by electric charge (d) None of the above

343. Hydrates are defined as---------
(a) Compounds with water molecules attached to them
(b) Compounds with their water molecules removed
(c) Compounds that have been heated to high temperatures
(d) None of the above

344. The difference between dipole-dipole forces and hydrogen bonds are that---------
(a) Dipole-dipole forces only exist between non polar molecules
(b) Dipole-dipole forces occur between polar molecules
(c) Dipole-dipole forces are caused by the interaction of partial charges on both molecules.
(d) None of the above are able to distinguish between dipole-dipole forces and hydrogen bonds

345. The main difference between a suspension and a colloid is that------------
(a) In suspensions the particles eventually settle to the bottom
(b) In colloids the particles eventually settle to the bottom
(c) In colloids, the solute is permanently dissolved in the solvent
(d) None of the above

346. Which of the following gases is unsuitable for use as a GC carrier gas?
(a) Nitrogen (b) Helium (c) Oxygen (d) All of the above

347. Which of these effects results from slow injection of a large sample volume?
(a) Increased resolution (b) Decreased resolution
(c) Non-linear detector response (d) None of the above

348. What information can be gained from running a gas chromatogram?
(a) Number of compounds present in the mixture
(b) Percent composition
(c) Retention time
(d) Molecular weight

349. What is the role of the helium gas when performing gas chromatography?
(a) It acts as the stationary phase
(b) It acts as the mobile phase
(c) It doesn't play a role in gas chromatography
(d) None of the above

350. ---------------- is the pigment present in tomato.
(a) Lycopene (b) Chlorophyll (c) Xanthophyll (d) Carotene

351. ------------- is also known as anti-egg white injury factor.
(a) Avidin (b) Ovalbumin (c) Ovoglobin (d) Biotin

352. A gene sequence under the coordinated control of a single operator is called----------
(a) Structural gene (b) Regulator gene (c) Operon (d) Repressor

353. The major pigment from turmeric is--------------
(a) Capsanthin (b) Citral (c) Curcumin (d) Lycopene

354. Which of the following spice contain eugenol?
(a) Clove (b) Cinnamon (c) Coriander (d) Capsicum

355. What is the term for nutrients that have been used by our body?
(a) Catalysts (b) Metabolites (c) Metabolism (d) Enzymes

356. Which of the following vitamins enhances the anticancer effect of garlic?
(a) Vitamin A (b) Vitamin K (c) Vitamin C (d) Vitamin D

357. Which of the following would be considered a disease of aging?
(a) Asthma (b) Cystic fibrosis (c) Type 1 diabetes (d) Cardiovascular disease

358. Which of the following food is a source of phytoestrogens?
(a) Soybeans and soy products (b) Coffee beans
(c) Dark green leafy vegetables (d) Black and green tea

359. For acetic acid production, the methods followed are-------------
(a) Orleans process (b) Rapid process (c) Submerged process (d) All of the above

360. The raw material for citric acid production is--------------
(a) Corn (b) Molasses (c) Starch (d) None of the above

361. Which of the following is not an amino acid?
(a) Glutamic acid (b) Aspartic acid (c) Glutamine (d) Palmitic acid

362. What type of covalent bonds link the amino acids in a protein?
(a) Peptide bonds (b) Hydrogen bonds (c) Ionic bonds (d) Ester bonds

363. Which of the following is an amino acid that is found in proteins?
(a) Adenosine (b) Adenine (c) Alanine (d) Linoleic acid

364. Which of the following factors can affect enzyme activity?
(a) Temperature (b) pH
(c) The presence of certain metal ions (d) All of the above

365. Glycolysis is the name given to the pathway involving the conversion of--------------
(a) Glycogen to glucose-6-phosphate
(b) Glycogen or glucose to fructose
(c) Glycogen or glucose to pyruvate or lactate
(d) Glycogen or glucose to pyruvate or acetyl CoA

366. The synthesis of glucose from lactate, glycerol, or amino acids is called as-------------
(a) Glycogenolysis (b) Glycolysis (c) Lipolysis (d) Gluconeogenesis

367. The major source of carbohydrate in a typical Western diet is-------------
(a) Starch (b) Cellulose (c) Sucrose (d) Saccharin

368. Liver glycogen breakdown is stimulated by-------------
(a) Insulin (b) Glucagon (c) Adrenaline (d) Both b and c

369. The process of breaking down triacyl glycerol into free fatty acids and glycerol is called-----
(a) Beta oxidation (b) Lipogenesis (c) Lipolysis (d) None of the above

370. Which of the following nucleotide bases is not found in RNA?
(a) Thymine (b) Adenine (c) Uracil (d) Guanine

371. Which of the following molecules does not form part of DNA?
(a) Purine (b) Pyrimidine (c) Deoxyribose (d) Amino acid

372. Which of the following is not a dietary antioxidant?
(a) Vitamin C (b) Lipoic acid (c) Vitamin K (d) Vitamin E

373. Calcium----------------

(a) Abates rheumatism

(b) Enhances strong bones

(c) Encourages the growth of beneficial bacteria in the gut

(d) None of the above

374. Which of the following statements regarding macronutrient energy value is TRUE ?

(a) Carbohydrates provide 9kcal/g consumed (b) Protein provides 2 kcal/g consumed

(c) Fat provides 7kcal/g consumed (d) Protein provides 4 kcal/g consumed

375. Which of the following statements regarding fiber is TRUE?

(a) Insoluble fiber has been shown to reduce total cholesterol

(b) Insoluble fiber has been shown to improve blood sugar control in diabetics

(c) Soluble fiber enhances gastric emptying

(d) Soluble fibers can also help lower LDL blood cholesterol by interfering with the absorption of dietary cholesterol

376. Which of the following values represent the recommended daily intake of fiber per day?

(a) 35-40g (b) 25-30g (c) 15-20g (d) 10-15g

377. The nitrogen bases which pair with two hydrogen bonds are-------------

(a) Adenine and thymine (b) Adenine and cytosine

(c) Cytosine and guanine (d) Cytosine and adenine

378. DNA differs from RNA in--------------

(a) Presence of deoxyribose sugar (b) Presence of thymine base

(c) Property of replication (d) All of the above

379. What is the term for nutrients that have been used by our body?

(a) Catalysts (b) Metabolites (c) Metabolism (d) Enzymes

380. Excess ----------------in the human body are stored in the adipose tissue.

(a) Fats (b) Proteins (c) Vitamins (d) Minerals

381. ---------------- is the probiotic present in milk.

(a) *E.coli* (b) *Lactobacillus* (c) *Alkaligenes* (d) *Bacillus* spp

382. Excess glucose in the human body are stored as ------------- in the liver.

(a) Glycogen (b) Glucose (c) Glycerol (d) Glycerine

383. Which acid is secreted by the cells of the gastric glands in the stomach?

(a) Hydrochloric acid (b) Sulphuric acid

(c) Nitric acid (d) Hydroiodic acid

384. Which organ secretes a chemical compound, which emulsifies fats and makes the food alkaline?

(a) Liver (b) Pancreas (c) Stomach (d) Large intestine

385. Lack of ---------- in the human body causes a severe deficiency disease called Kwashiorkor.

(a) Proteins (b) Carbohydrates (c) Fats (d) Lipids

386. In which part of the body fats, carbohydrates and proteins are digested?
(a) Small intestine (b) Large intestine (c) Stomach (d) Mouth

387. The bile duct carries bile from the liver to the --------------
(a) Small intestine (b) Large intestine (c) Pancreas (d) Stomach

388. Which of the following is not a simple sugar (monosaccharide)?
(a) Sucrose (b) Glucose (c) Fructose (d) Galactose

389. The leading cause of blindness in the world today is a dietary deficiency of --------------
(a) Vitamin A (b) Vitamin D (c) Vitamin E (d) Vitamin K

390. Other than phosphocreatine breakdown, the fastest way to resupply ATP to a muscle is------
(a) Breakdown fat stores in muscle and convert fatty acids to glucose
(b) Anaerobic glycolysis
(c) Aerobic glycolysis
(d) To stop and eat a meal rich in carbohydrate

391. Energy from the metabolism of energy nutrients is released as ----------------
(a) Heat (b) ATP (c) Water (d) Both a and b

392. Low density lipoproteins are the principle transport vehicle for --------- in the blood.
(a) Cholesterol (b) Phospholipids (c) Glucose (d) Triglycerides

393. Thiamin, niacin, and riboflavin work together in important biochemical pathways that -------
(a) Release energy from carbohydrate, fat, and protein
(b) Control visual process
(c) Synthesize collagen
(d) Promote absorption of calcium

394. Factors such as heat, iron, copper and oxygen ---------------
(a) Are necessary to convert vitamin C from its inactive to active form
(b) Destroy vitamin C in foods
(c) Enhance the stability of vitamin C in foods
(d) Have no effect on vitamin C

395. The principle form of lipid in food is---------------
(a) Triglyceride (b) Fiber (c) Starch (d) Fatty acid

396. The fatty acids in margarine and shortenings have a relatively high concentration of---------
(a) Propanoic acid (b) Trans fatty acids (c) cis fatty acids (d) Omega-3 fatty acids

397. Body can manufacture the blood sugar, glucose from-------------
(a) Alcohol (b) Fatty acids (c) Fibers (d) Protein

398. The difference between fats and oils is, fats------------
(a) Have a high percentage of fatty acids as saturated fatty acids
(b) Have a high percentage of fatty acids as unsaturated fatty acids

(c) Are solid at room temperature

(d) Both a and c

399. The protein-digesting enzymes, trypsin, chymotrypsin and carboxypeptidase, are produced by the-------------

(a) Pancreas (b) Liver (c) Duodenum (d) Stomach

400. If a fatty acid has one double covalent (carbon-carbon) bond it is -------------

(a) An omega-1 fatty acid (b) A monounsaturated fatty acid

(c) A polyunsaturated fatty acid (d) A saturated fatty acid

401. People who eat fish, especially fish with abundant EPA and DHA, appear to have a lower risk for heart disease. The reason for this benefit appears to be ---------------

(a) The type of eicosanoids synthesized in the body

(b) The metabolic pathways that favor blood clotting mechanisms

(c) A decrease in formation of blood clots

(d) Both a and b

402. In the body, excess energy is-------------

(a) Stored as free fatty acids in muscle tissue

(b) Stored as glycogen, as the body has unlimited capacity to store water and carbohydrate

(c) Stored as triglycerides in adipose tissue

(d) Excreted by way of the urine and stool

403. Osteoporosis is---------------

(a) Caused by a dietary lack of vitamin D

(b) Indicates the person has been taking certain medications that interfere with normal bone metabolism

(c) A general term for a disease that results in reduced bone mass

(d) Is another name for osteomalacia

404. A diet high in fat, low in dietary fiber, and excess food energy may be an important contributing factor in the development of -------------

(a) Phytobezoars (b) Dental caries

(c) Lactose intolerance (d) Colon cancer

405. The RDA's are based on-------------

(a) Average requirements for a population

(b) Precise numbers derived from carefully controlled experiments

(c) Average needs, plus increased amounts sufficient to meet requirements for nearly all variability in the population

(d) The fact that every nutrient must be consumed every day

406. The human brain and other nerve tissue use mostly ---------------- as fuel.

(a) Galactose (b) Glycogen (c) Fructose (d) Glucose

407. In preserving vitamins in foods, which of the following techniques is most effective?

(a) Cut fruits and vegetables into as small portions as possible so they are easy to eat and the nutrients are more available to the digestive processes

(b) Discard the outer leaves and skins of vegetables and fruits; the nutrients are tucked into the inner tender portions of produce

(c) Keep fruits and vegetables at room temperature since that is the temperature at which they were grown

(d) Cook fruits and vegetables in as little water as possible and for a brief period of time so they are tender crisp

408. Omega-3 and omega-6 fatty acids are designated-------------fatty acids.

(a) Trans (b) Saturated (c) Dietary essential (d) Non essential

409. Green leafy vegetables are an excellent source of---------------

(a) Riboflavin (b) Vitamin K (c) Beta carotene (d) All of the above

410. A difference between liver glycogen and muscle glycogen is -------------

(a) Muscle glycogen is linked by -glycosidic linkages so it cannot be quickly broken down to release glucose while liver glycogen is linked by -linkages

(b) Liver glycogen is an important source of blood glucose, while muscle glycogen is not

(c) Liver glycogen is predominantly amylopectin, while muscle glycogen is predominantly amylose

(d) Liver glycogen is predominantly amylose while muscle glycogen in predominately amylopectin

411. A clinical symptom of iron-deficiency anemia is-------------

(a) Intolerance to cold temperature (b) Pale skin

(c) Fatigue (d) All of the above

412. Beta carotene is also called as ---------------

(a) Chlorophyll (b) Retinoic acid, retinal or retinal

(c) Calcitriol (d) Provitamin A

413. Spinach has a very high iron content, but low bioavailability. The iron in spinach is in the form of---------------

(a) Elemental iron (b) Myoglobin (c) Nonheme iron (d) Hemoglobin

414. Many green, leafy vegetables are high in dietary calcium, but the calcium is unavailable because -----------

(a) It is absorbed and then reexcreted through the small intestine

(b) The calcium is complexed to phytic acid so it can't be absorbed

(c) Bile acids carry the calcium out of the body by way of the feces

(d) The calcium is complexed to oxalic acid so it can't be absorbed

415. Which of the following is not a reason that additives are put into food?

(a) Promoting good health (b) Increasing cost

(c) Improving texture and flavour (d) Making food more convenient

416. Yogurt is an appropriate substitute for milk in the lactose intolerant individual because-------

(a) The lactose in yogurt is better tolerated

(b) The manufacturers have removed all the lactose before processing

(c) Lactase is added to the milk before it is sold

(d) The bacteria that converts milk to yogurt digests the lactose

417. A deficiency of vitamin E is most likely found in------------

(a) Infants and small children

(b) People with a very high intake of foods rich in PUFA

(c) Vegetarians

(d) People with fat malabsorption syndromes

418. All B vitamins function as ---------------

(a) Sources of energy (b) Coenzymes

(c) Cofactors (d) Enzymes

419. MSG is-------------

(a) A natural protein

(b) A flavouring agent

(c) A flavour enhancer, bringing out the natural flavour of a food

(d) An extender, adding texture and bulk to a product

420. Nutritionally, the difference between brown rice and white rice is---------------

(a) The presence of bran and germ layers in the brown rice, which has most of the nutrients

(b) Negligible. There is a little more thiamin in white rice than brown rice

(c) Fiber content. White rice has more fiber.

(d) Both a and c

421. Most alcohol is metabolized in----------------

(a) The brain (b) The walls of the small intestine

(c) The liver (d) The stomach

422. Vitamin B_{12} is --------------

(a) Not a true vitamin, but rather a hormone (b) Present only in cereal grains

(c) Present only in foods of animal origin (d) Present only in fruits and vegetables

423. Ice cream with no additives is likely to -------------- than ice cream with additives.

(a) Have more added colourings (b). Weigh less

(c) Cost less (d) Cost more

424. In the industrial production of streptomycin, the secondary metabolite is-------------

(a) Vitamin B_{12} (b) b. Vitamin C (c) Vitamin B_6 (d) Ethanol

425. Which of the following is a disaccharide?

(a) Sucrose (b) Galactose (c) Fructose (d) Glucose

426. Muscles store metabolic reserves as---------------

(a) Carbohydrates (b) Glycogen (c) Amino acids (d) Triglycerides

427. Which food item is used as an additive in shoe polish?

(a) Amyl acetate (b) Shellac (c) Corn (d) Sugar

428. The liver utilizes fatty acids to synthesize-----------

(a) Triglycerides (b) Steroids (c) Phospholipids (d) All of the above

429. Beta-oxidation----------------

(a) Occurs in the mitochondria

(b) Is the process that breaks down fatty acids into two-carbon fragments that can be metabolized by the TCA cycle

(c) Yields large amounts of ATP, while requiring coenzyme A, NAD, and FAD

(d) None of the above

430. Lipoproteins that carry mostly cholesterol and phospholipids from peripheral tissues to the liver are called------------

(a) Low-density lipoproteins (LDLs) (b) High-density lipoproteins (HDLs)

(c) Both a and b (d) None of the above

431. Diets containing too many calories and too many lipids by proportion increase the incidence of ----------

(a) Obesity (b) Heart disease and atherosclerosis

(c) Diabetes (d) All of the above

432. Which of the following is a complete protein source?

(a) Eggs (b) Corn (c) Rice (d) None of the above

433. In transamination, the amino group of an amino acid is--------------

(a) Converted to ammonia

(b) Converted to urea

(c) Transferred to another carbon chain

(d) Transferred to a molecule in the glycolytic pathway

434. The process of deamination produces--------------

(a) Keto acids (b) Urea (c) Ammonia (d) Acetyl CoA

435. Nitrogen compounds of the body include--------------

(a) Amino acids (b) Nucleotides (c) Creatine (d) All of the above

436. B vitamins are --------------

(a) Important to many metabolic reactions (b) Involved in the oxidation of food

(c) Water soluble (d) All of the above

437. Much of what we consider flavour is actually detected by the-------------

(a) Nose (b) Tongue (c) Eyes (d) Stomach

438. A cofactor that is essential for muscle contraction, nerve function, and blood clotting is-----
(a) Sodium (b) Potassium (c) Calcium (d) Magnesium

439. Which food group is the best source of vitamin C?
(a) Dairy (b) Meats
(c) Fruits and vegetables (d) Desserts

440. A mineral that is a necessary cofactor for hemoglobin synthesis is----------------
(a) Zinc (b) Copper (c) Cobalt (d) Iodine

441. Which of the following is a trace element?
(a) Chlorine (b) Sodium (c) Calcium (d) Iron

442. Natural flavourings are processed in-------------
(a) Farms (b) Factories (c) India (d) Laboratories

443. Which food group is the best source of calcium?
(a) Breads (b) Meats (c) Dairy (d) Fruits

444. Impaired fat absorption in the intestine would interfere with the absorption of which vitamin?
(a) Vitamin A (b) Vitamin B_{12} (c) Vitamin C (d) Riboflavin

445. A mineral necessary for normal membrane function is-------------
(a) Calcium (b) Sodium (c) Manganese (d) Zinc

446. Factors that influence an individual's BMR include-------------
(a) Sex (b) Age (c) Body weight (d) All of the above

447. Omega 3 fatty acids------------
(a) Improve mental balance (b) Prevent anaemia
(c) Boost heart health and lower triglycerides (d) None of the above

448. Folic acid-------------
(a) Eases digestion
(b) Reduces the risk of neural tube defects with unborn children
(c) Improves the power of concentration
(d) None of the above

449. Lycopene--------------------
(a) Prevents thyroid gland diseases
(b) Abates headaches
(c) Has been found to possess antioxidant and antiproliferative properties
(d) None of the above

450. Dietary fibres----------------
(a) Improve sight (b) Reduce the risk of heart diseases
(c) Reduce the risk of colon cancer (d) None of the above

451. The raw materials from botanic sources used in the flavour and fragrance industries are known as---------

(a) Oleoresins (b) Essential oils (c) Flavourings (d) All of the above

452. Deterioration of fat in the presence of oxygen is due to the enzyme-------------

(a) Lipase (b) Protease (c) Lipoxygenase (d) Peptidase

453. In cheese manufacture, curd is formed by the addition of---------------

(a) Rennin (b) Protease (c) Peptidase (d) Amylase

454. Excessive intake of polished rice causes deficiency of--------------

(a) Vitamin A (b) Vitamin B_1 (c) Vitamin D (d) Vitamin K

455. A toxic element which may contaminate food is ----------------

(a) Zinc (b) Copper (c) Lead (d) Calcium

456. The toxin found in groundnut kernel is---------------

(a) Aflatoxin (b) Patulin (c) Aminotoxin (d) Paratoxin

457. Decomposition of carbohydrates by microorganisms or enzymes is called as-----------

(a) Putrefaction (b) Fermentation (c) Canning (d) Dextrinisation

458. The toxic substance gossypol is found in-----------------

(a) Rapeseed (b) Linseed (c) Cotton (d) Fababean

459. Amino acids are used as food additives for which of the following reasons?

(a) As natural antibiotics (b) As natural growth inhibitors

(c) For nutritive purposes (d) As antioxidants

460. Lycopene is an important antioxidant normally present in high level in-------------

(a) Artichokes (b) Bananas (c) Tomatoes (d) Soybeans

461. Fats and oils have an important function in food preparation. They can-----------

(a) Sweeten a food product (b) Assist some foods to retain moisture

(c) Assist in binding ingredients in a cake (d) Contribute to the aeration of food products

462. Invert sugar is-------------

(a) Lactose (b) Sucrose

(c) Hydrolytic products of sucrose (d) Fructose

463. The monosaccharide units are linked by 1 → 4 glycosidic linkage in------------

(a) Maltose (b) Sucrose (c) Cellulose (d) Cellobiose

464. Which of the following is a non-reducing sugars?

(a) Isomaltose (b) Maltose (c) Lactose (d) Trehalose

465. Mutarotation refers to change in-------------

(a) pH (b) Optical rotation (c) Conductance (d) Chemical properties

466. A polysaccharide which is often called animal starch is---------------

(a) Glycogen (b) Starch (c) Inulin (d) Dextrin

467. Starch is a-------------

(a) Polysaccharide (b) Monosaccharide (c) Disaccharide (d) None of the above

468. The most abundant carbohydrate found in nature is-------------

(a) Starch (b) Glycogen (c) Cellulose (d) Chitin

469. Maltose can be formed by hydrolysis of---------------

(a) Starch (b) Dextrin (c) Glycogen (d) All of the above

470. A carbohydrate commonly known as invert sugar is------------

(a) Fructose (b) Sucrose (c) Glucose (d) Lactose

471. A heteropolysaccharide among the following is-------------

(a) Inulin (b) Cellulose (c) Heparin (d) Dextrin

472. Consumption of iodized salt is recommended for prevention of-------------

(a) Hypertension (b) Hyperthyroidism (c) Endemic goiter (d) None of the above

473. Restriction of salt intake is generally recommended in-----------

(a) Diabetes mellitus (b) Hypertension (c) Cirrhosis of liver (d) Peptic ulcer

474. β-amylase is------------

(a) Endoenzyme (b) Exoenzyme

(c) Saccharifying enzyme (d) Both b and c

475. Juice clarification is facilitated by using-------------

(a) Cellulases (b) Inulinase (c) Amylase (d) Lactase

476. Lysozyme is naturally present in--------

(a) Egg white (b) Bacteria (c) Tears and milk (d) All of the above

477. Enzymes act as antimicrobials by----------------

(a) Depriving an organism of a necessary metabolite

(b) Generating a substances toxic to the organism

(c) Attracting a cell wall component

(d) All of the above

478. Trichoderma β-glucanase is reported to---------------

(a) Stabilize mashing

(b) Convert taste fractions of dextrins to fermentable sugars in beer

(c) Convert starch to dextrin

(d) All of the above

479. An ion exchange resin is composed of--------------

(a) Polymeric network (b) Ionic functional groups

(c) Counter ions (d) All of the above

480. Which medium is used for the production of penicillin using immobilized cells-----------

(a) 1% peptone (b) Glucose (c) Yeast extract (d) LB broth

481. The immobilized enzyme produced by micro encapsulation technique provides------------
(a) An extremely large surface area (b) Smaller surface area
(c) High amount of solvent (d) Relatively smaller surface area

482. The chemical oxygen demand measures the--------------
(a) Amount of oxygen required for growth of microorganisms in water
(b) Amount of oxygen that would be removed from the water in order to oxidize pollution
(c) Amount of oxygen required to oxidize the calcium present in waste water
(d) None of the above

483. The active site of an enzyme remains---------------
(a) At the center of globular proteins (b) Rigid and does not change shape
(c) Complementary to the rest of the molecule (d) None of the above

484. The enzyme inhibition can occur by---------------
(a) Reversible inhibitors (b) Irreversible inhibitors
(c) Both a and b (d) None of the above

485. The bitter taste of the high protein materials is reduced by using--------------
(a) Invertase (b) Dectinase (c) Protease (d) None of the above

486. Sulfhydryl oxidase is used for-------------
(a) Discoloration (b) Clarification of images
(c) Removal of off flavour in UHT milk (d) All of the above

487. Liquefaction of starch to dextrin is carried out by--------------
(a) α-amylase (b) Cellulase (c) Pectinase (d) All of the above

488. Which is true about rennet?
(a) It is a mixture of protease chymosin and pepsin
(b) It is a mixture of rennin and pepsin
(c) Both a and b
(d) None of the above

489. Milk digestibility is improved by using--------------
(a) Rnase (b) Lactase (c) β-amylase (d) None of the above

490. Which of the following mainly serve to convert starch into high fructose corn syrup?
(a) α-amylase (b) Gluco-isomerase (c) Gluco-amylase (d) All of the above

491. Hesperidinase is used for---------------
(a) Juice clarification (b) Juice debittering
(c) Off flavour reduction (d) RNA reduction in fish

492. The enzyme used to reduce bitterness of grapes commonly contains---------------
(a) α-L-rhamonosidase (b) β-d-glucosidase
(c) β- galactosidase (d) Both a and b

493. Citrus juice debittering can be carried out using-------------
(a) Limoninase (b) Inulinase (c) Anthocyanase (d) None of the above

494. Which of the following enzyme is responsible for causing vitamin B deficiency disease beriberi?
(a) Ascorbic acid oxidase (b) Thiaminase
(c) Lipoxygenase (d) None of the above

495. Off flavour removal in soya may be achieved by using-------------
(a) Di acetyl reductase (b) β- amylase
(c) Aldehyde oxidase (d) Protease

496. The reduction in off flavour of beer is practiced through-------------
(a) Hersperidinase (b) R nase (c) Diacetyl reductase (d) Invertase

497. Removal of tannins from fruit juices can be achieved by treatment with--------------
(a) Polyvinyl pyrrolidone (b) Bentonite
(c) Pectinol (d) Papain

498. Alum is used in purification of water to-----------------
(a) Kill pathogens (b) Kill non pathogens
(c) Remove organic matter (d) None of the above

499. The enzyme used in wine and beer industry to avoid haze formation is---------------
(a) Protease (b) Amylase (c) Raffinase (d) Isomerase

500. The most abundant unsaturated fatty acids in cow milk fat is-------------
(a) Oleic (b) Linoleic (c) Searic (d) Pamitic

501. Which is the major fraction of maize protein?
(a) Glutelin (b) Gliadin (c) Gluten (d) None of the above

502. Which is a effective antimutagen phenolic constituent present in strawberries?
(a) Tannin (b) Ellagic acid (c) Citric acid (d) Anthocyanin

503. Which acid is produced as a result of activity of insects and rodents during the storage of farm produce?
(a) Oxalic acid (b) Uric acid (c) Citric acid (d) Malic acid

504. Which of the following is not a disinfectant containing a heavy metal?
(a) Silver nitrate (b) Mercurochrome (c) Chlorine (d) Copper sulfate

505. Cations have-------------
(a) Positive charge (b) Negative charge
(c) No charge (d) It is impossible to predict the charge on a cation

506. The main difference between a suspension and a colloid is that-------------
(a) In suspensions the particles eventually settle to the bottom
(b) In colloids the particles eventually settle to the bottom
(c) In colloids, the solute is permanently dissolved in the solvent
(d) None of the above

507. An unsaturated solution----------------
(a) Hasn't dissolved as much solute as is theoretically possible
(b) Has dissolved exactly as much solute as is theoretically possible
(c) Is unstable because it has dissolved more solute than would be expected
(d) None of the above

508. Sodium nitrate is-------------
(a) Anticaking agent (b) Antioxidant (c) Curing agent (d) Colourant

509. Mirror image stereoisomer's are called as-------------
(a) Enantiomers (b) Diastereoisomers (c) Isostereomers (d) cis stereomers

510. Linear inhibition is sometimes called as---------------
(a) Complete inhibition (b) Incomplete inhibition
(c) Partial inhibition (d) Mixed inhibition

511. Yield coefficient represents----------------
(a) Total biomass or product produced
(b) Conversion efficiency of a substrate into product
(c) Conversion rate of a substrate into biomass or product
(d) Production time of biomass or product

512. Chymosin hydrolyses the bond between--------------
(a) Alanine and glycine (b) Phenyl alanine and methonine
(c) Glutamic acid and alanine (d) Alanine and phenyl alanine

513. White flour that has missing vitamins and minerals added back is called ------------- flour.
(a) Enriched (b) All-natural (c) Fortified (d) Whole wheat

514. Flavonoids are chemicals mainly found in------------
(a) Milk and yogurt (b) Oil and sugar (c) Meat and beans (d) Fruits and vegetables

515. The daily recommendations for dietary fiber are based on---------------
(a) Altitude and lung capacity (b) Activity level and weight
(c) Age and gender (d) Height and weight

516. The major fibrous proteins are--------------
(a) Peptone and edestin (b) Glutelin and leucine
(c) Valine and lysine (d) Myosin and actin

517. Which of the following organs of the body produces insulin?
(a) Liver (b) Stomach (c) Pancreas (d) Small intestine

518. In ion-exchange chromatography ,proteins are separated on the basis of their -------------
(a) Net charge (b) Size (c) Shape (d) Either b or c

519. Which of the following may be added to stabilize the protein after yeast cells disruption?
(a) NaCl (b) Protease inhibitor (c) AMP (d) All of the above

520. Gel-filtration chromatography separates on the basis of---------
(a) Size and shape using porous beads packed in a column
(b) Size using porous beads packed in a column
(c) Shape using porous beads packed in a column
(d) None of the above

521. Affinity chromatography deals with the----------
(a) Specific binding of a protein constituents for another molecule
(b) Protein - protein interaction
(c) Protein - carbohydrate interaction
(d) None of the above

522. A purified protein sample contains 10 μg of protein and has an enzyme activity of 1 m mole of ATP synthesized/sec (1 unit). What is the specific activity of the final purified sample?
(a) 1,000 units/mg (b) 10,000 units/mg (c) 100,000 units/mg (d) 1,000,000 units/mg

523. Proteins separation can be carried out on the basis of----------
(a) Net charge (b) Solubility in salt solutions
(c) Size or mass (d) All of the above

524. Best way to determine the location of protein in the purification scheme is to measure the---
(a) Rate of ATP synthesis (b) Changes in the refractive index
(c) UV absorption (d) Mass spectroscopy of the protein

525. Which of the following chemicals is a solvent used for estimation of fat content?
(a) Hydrochloric acid (b) Sodium hydroxide
(c) Amylase (d) Ether

526. Which of the following analytical methods can be used to distinguish flavour compounds?
(a) Hydrometry (b) Near infrared spectroscopy
(c) Polarimetry (d) Gas chromatography

527. Table sugar is which type of sugar?
(a) Fructose (b) Galactose (c) Glucose (d) Sucrose

528. An additive that keeps moist foods from drying out is called as---------------
(a) Preservative (b) Stabilizer (c) Humectant (d) Emuslifier

529. Emulsifiers are used to do what?
(a) Enhance mouthfeel (b) Keep oil and water mixed
(c) Add brown or golden colouring (d) Prevent food from spoiling

530. What additive in salt prevents the goiter?
(a) Sodium caseinate (b) Titanium oxide
(c) Cochineal extract (d) Potassium iodide

ANSWER KEY

1. d	2. a	3. a	4. a	5. b	6. a	7. b	8. b	9. a	10. a
11. a	12. a	13. b	14. a	15. c	16. a	17. c	18. b	19. a	20. a
21. a	22. a	23. b	24. c	25. b	26. c	27. a	28. c	29. d	30. d
31. a	32. b	33. d	34. a	35. a	36. d	37. c	38. b	39. d	40. c
41. b	42. c	43. c	44. c	45. b	46. a	47. a	48. b	49. b	50. d
51. c	52. c	53. d	54. c	55. a	56. c	57. d	58. b	59. b	60. c
61. c	62. c	63. b	64. c	65. a	66. b	67. d	68. b	69. a	70. c
71. d	72. c	73. a	74. b	75. b	76. a	77. c	78. a	79. b	80. d
81. a	82. d	83. b	84. d	85. a	86. a	87. a	88. d	89. c	90. a
91. b	92. c	93. b	94. d	95. a	96. c	97. a	98. b	99. a	100. c
101. b	102. a	103. a	104. a	105. b	106. a	107. c	108. c	109. b	110. d
111. a	112. b	113. b	114. c	115. c	116. b	117. a	118. b	119. a	120. a
121. b	122. a	123. b	124. a	125. a	126. a	127. c	128. d	129. a	130. b
131. a	132. a	133. d	134. a	135. a	136. b	137. c	138. a	139. c	140. a
141. a	142. d	143. b	144. b	145. b	146. b	147. c	148. a	149. c	150. c
151. a	152. d	153. b	154. b	155. c	156. b	157. a	158. b	159. c	160. a
161. b	162. b	163. c	164. d	165. d	166. a	167. c	168. a	169. c	170. c
171. d	172. b	173. a	174. b	175. a	176. b	177. b	178. c	179. a	180. b
181. d	182. b	183. c	184. d	185. d	186. a	187. d	188. c	189. a	190. c
191. b	192. b	193. a	194. a	195. c	196. c	197. b	198. c	199. a	200. c
201. b	202. a	203. a	204. b	205. a	206. b	207. b	208. b	209. b	210. b
211. b	212. c	213. a	214. a	215. a	216. d	217. d	218. a	219. d	220. d
221. b	222. b	223. b	224. b	225. b	226. a	227. c	228. d	229. b	230. c
231. b	232. a	233. c	234. b	235. a	236. c	237. d	238. d	239. a	240. d
241. a	242. b	243. a	244. b	245. c	246. d	247. b	248. b	249. d	250. a
251. b	252. b	253. c	254. c	255. a	256. b	257. a	258. b	259. a	260. c
261. a	262. a	263. c	264. c	265. b	266. d	267. a	268. c	269. a	270. b
271. b	272. c	273. d	274. b	275. b	276. b	277. a	278. c	279. a	280. b
281. c	282. b	283. a	284. a	285. d	286. b	287. d	288. a	289. a	290. d
291. b	292. c	293. b	294. a	295. b	296. a	297. a	298. a	299. d	300. a
301. a	302. a	303. b	304. a	305. b	306. b	307. d	308. b	309. a	310. d
311. c	312. b	313. a	314. a	315. c	316. a	317. a	318. b	319. a	320. a
321. d	322. b	323. a	324. a	325. a	326. a	327. d	328. c	329. a	330. d
331. d	332. a	333. c	334. b	335. c	336. b	337. a	338. d	339. d	340. a
341. b	342. c	343. a	344. d	345. a	346. c	347. b	348. a	349. b	350. a
351. d	352. c	353. c	354. a	355. b	356. a	357. d	358. a	359. d	360. a
361. d	362. a	363. c	364. d	365. c	366. d	367. a	368. d	369. c	370. a

371. d	372. c	373. b	374. a	375. d	376. a	377. a	378. d	379. b	380. a
381. b	382. a	383. a	384. a	385. a	386. a	387. a	388. a	389. a	390. b
391. d	392. a	393. a	394. b	395. a	396. b	397. d	398. d	399. a	400. b
401. d	402. c	403. c	404. d	405. c	406. d	407. d	408. c	409. d	410. b
411. d	412. d	413. c	414. d	415. b	416. d	417. d	418. b	419. c	420. a
421. c	422. c	423. d	424. a	425. a	426. b	427. c	428. d	429. d	430. b
431. d	432. a	433. c	434. c	435. d	436. d	437. a	438. c	439. c	440. b
441. d	442. b	443. c	444. a	445. b	446. d	447. c	448. b	449. c	450. c
451. b	452. c	453. a	454. b	455. c	456. a	457. b	458. c	459. d	460. c
461. d	462. c	463. a	464. d	465. b	466. d	467. a	468. c	469. d	470. b
471. c	472. c	473. b	474. d	475. a	476 d	477. d	478. a	479. d	480. b
481. a	482. b	483. d	484. c	485. c	486. c	487. a	488. c	489. b	490. d
491. a	492. d	493. a	494. b	495. c	496. c	497. a	498. c	499. a	500. a
501. a	502. b	503. b	504. c	505. a	506. a	507. a	508. c	509. a	510. a
511. b	512. b	513. a	514. d	515. c	516. d	517. c	518. a	519. d	520. a
521. a	522. c	523. d	524. a	525. d	526. d	527. d	528. c	529. b	530. d

4

Food Microbiology and Safety

1. The toxin produced by some strains of the fungus *Aspergillus flavus* is----------------
(a) Aflatoxin (b) Mycotoxin (c) Enterotoxin (d) Exotoxin
2. Enzymatic anaerobic decomposition of proteins, peptide or amino acid is termed as------------
(a) Putrefaction (b) Rancidity (c) Decay (d) Deproteination
3. ------------- rot is a microbial spoilage caused by *Ervionnia crarotovora* which results in soft mushy consistency and bad odour.
(a) Bacterial (b) Fungal (c) Erwinia (d) Gaint
4. The term used to describe the metabolism of organism in food containing protein and carbohydrate is known as----------------
(a) Protein sparing (b) Protein metabolism (c) Protein synthesis (d) None of the above
5. The isolation of desired microorganisms from its natural habitats is known as--------------
(a) Screening (b) Search (c) Investigation (d) None of the above
6. A type of bioreactor used to obtain desired product from microorganisms is---------------
(a) Fermentor (b) Chemostat (c) Turbidostat (d) All of the above
7. ------------ are the secondary metabolites obtained from microorganisms and used to kill other microorganisms.
(a) Enzymes (b) Proteins (c) Antibiotics (d) Drugs
8. ------------- are the low molecular weight protein or protein complexes produced by bacteria that inhibit or kill other bacteria.
(a) Antibiotics (b) Bacteriocins (c) Drugs (d) None of the above
9. Fermentation in which two or more than two microorganisms are utilized is called as-----------
(a) Continuous fermentation (b) Anaerobic fermentation
(c) Dual fermentation (d) All of the above
10. --------------is the term referred for addition of microbial culture inside the fermentor.
(a) Media pouring (b) Culturing (c) Inoculation (d) Microbe addition
11. The processes carried out after fermentation for purification of product is called as------------
(a) Upstream processes (b) Downstream processing
(c) Production processes (d) Past production steps

12. The type of fermentation in which microbial cells are growing throughout the fermentation media is known as -----------------fermentation.
(a) Surface (b) Solid state (c) Submerged (d) Continuous

13. --------------- is device used in fermentor to carry out the mixing of fermentation media.
(a) Agitator (b) Baffels (c) pH probe (d) Sparger

14. -------------- is the first reported algae used as SCP.
(a) BGA (b) Spirulina (c) Golden algae (d) Red algae

15. The fermented product obtained from cabbage is called as-------------
(a) Cabage pickle (b) Miso (c) Temphe (d) Sauerkraut

16. ------------ is the amount of free and combined available chlorine present in water considered during calculating chlorine demand.
(a) Residual chlorine (b) Chlorine dose
(c) Combined chlorine (d) Free chlorine

17. --------------yeast is mainly used in bakery.
(a) *Torulopsis* (b) *Saccharomyces cerevisiae*
(c) *Aspergillus* (d) None of the above

18. ---------------techniques are used for producing high yield of product.
(a) Strain improvement (b) Streak plate
(c) Biochemical tests (d) Screening

19. ----------- are the chemical compounds produced which offers the additional benefits to the microorganisms.
(a) Toxins (b) Secondary metabolites
(c) Metabolites (d) Catabolites

20. -------------------- refer to pathological condition results from ingestion of food or feeds contaminated with fungal toxins.
(a) Mycotoxicosis (b) Pathogenesis (c) Toxicosis (d) All of the above

21. -------------- are the anaerobic spore forming heat resistant bacteria present in canned foods.
(a) *Clostridium spp.* (b) *Bacillus spp.* (c) Yeast (d) *E.coli*

22. ---------------- is the probiotic present in milk.
(a) *E.coli* (b) *Lactobacillus* (c) *Alkaligenes* (d) *Bacillus* spp

23. ----------------- is the spoilage defect very common with bread.
(a) Ropiness (b) Black rot
(c) Hetero fermentation (d) Green rot

24. ----------------microorganism causes human listeriosis.
(a) *E.coli* (b) *Shigella*
(c) *Listeria monocytogenes* (d) *Alkaligenes*

25. Which enzyme causes sweet curdling in milk?
(a) Rennet (b) Caseinase (c) Proteinase (d) Lipases
26. Who is known as father of microbiology?
(a) Louis pasture (b) Edward Jenner
(c) A. Flemming (d) Antonj van –Leewenhoek
27. The conditions of pasteurization of milk are worked out based on destruction of -----------
(a) *Mycobacterium tuberculosis* (b) *Salmonella typhimurium*
(c) *Clostridium botulinum* (d) *Staphylococcus aureus*
28. Groundnut is prone to incidences of contamination with--------------
(a) *Aspergillus flavus* (b) *Aspergillus paraciticus*
(c) *Aspergillus niger* (d) Both a and b
29. The lowest temperature that kills all microorganisms in a liquid suspension in 10 minutes is known as the--------------
(a) Thermal death point (b) Thermal death temperature
(c) F value (d) D value
30. ------------- is the instrument used for sterilization, which uses steam under pressure.
(a) Hor air oven (b) Boiling water bath (c) Incubator (d) Autoclave
31. Which type of cell does not show the presence of nuclear membrane?
(a) Animal cell (b) Eukaryotes (c) Animal cell (d) None of the above
32. The energy generation in bacteria takes place in--------------
(a) Mitrochondria (b) Plasma membrane (c) Periplasmic space (d) Ribosomes
33. Phase of bacterial growth where the bacteria multiplies by exponential manner is known as--------------phase.
(a) Lag (b) Log (c) Stationary (d) Double
34. Temperature required to kill the bacteria in specific given time is known as------------
(a) TDT (b) TDP (c) D value (d) F value
35. --------- is the nutrient transport in bacteria uses the energy during the transport of nutrient across the cell wall.
(a) Active transport (b) Passive transport (c) Bulk transport (d) None of the above
36. ------------ is the external structure with bacterial cell wall which increases the virulence.
(a) Capsule (b) Flagella (c) Pili (d) Cell covering
37. Bacteria which multiply at high temperature are known as--------------
(a) Thermoduric (b) Thermophilic (c) Mesophilic (d) Thermosporic
38. Biogas is a mixture ------------ gases.
(a) CO_2 and CH_4 (b) CO_2 and NH_3 (c) NH_3 and CH_4 (d) CO and CH_4
39. -------- is the toxin formed as a result of the enzymatic modification of food in which *Bacillus cereus* is growing.
(a) Enterotoxin (b) Exotoxin (c) Emetic toxin (d) None of the above

40. Zearalenone toxin is produced by which fungi?
(a) *Fusarium spp* (b) *Aspergillus spp* (c) *Trichothecenes spp* (d) None of the above

41. Which of the following is a limitation of the autoclave?
(a) It lacks the ability to inactivate viruses (b) It lacks the ability to kill endospores
(c) It will destroy heat labile materials (d) It cannot be used with glassware

42. D value in death rate curve indicates time required for survival or death rate curve to traverses------------------
(a) One log cycle (b) 1/10 log cycle (c) 1/100 log cycle (d) None of the above

43. In canning Z value is defined as the temperature necessary to bring ---------- change in the TDT or D value ------------fold.
(a) 10 (b) 90 (c) 80 (d) 70

44. Which of the statement is true for continuous reactor at steady state?
(a) The rates of biomass, substrate and product concentrations are zero
(b) Biomass, substrate and product concentrations are zero
(c) Biomass, substrate and product concentrations do not change with time
(d) Biomass, substrate and product concentrations change with time

45. A continuous bioreactor in which only the flow rate is used to control the rate of cell or product productivity is called-------------
(a) Turbidostat (b) Chemostat (c) Level stat (d) pH

46. Who is the father of tissue culture?
(a) Roff Harrison (b) Moud Mentel (c) Broed Deake (d) Alexander Ruzvelt

47. The process of making beer is known as ------------------
(a) Winning (b) Brewing (c) Distillation (d) None of the above

48. Riboflavin is commercially produced by fermentation using which microorganisms?
(a) *Ashbya gossypii* (b) *Riboflavina spp.* (c) *Aspergillus gossypii* (d) None of the above

49. Which of the following is used to calculate mass of substrate in the reactor?
(a) Flow rate × substrate concentration in the reactor
(b) Flow rate × mass of reactor
(c) Volume of reactor × substrate concentration in reactor
(d) Volume of reactor × Flow rate

50. Buffers when used in fermentation media functions as-----------------
(a) Changes the pH (b) Resist the change in pH
(c) Changes pH towards acidic (d) Makes pH neutral

51. Which of the following is type of fungi?
(a) Yeast (b) Bacteria (c) Virus (d) None of the above

52. The rice beer or wine obtain from *Aspergillus oryzae* with 14-17% of alcohol is known as----
(a) Sake (b) Sherry (c) Dry wine (d) None of the above

53. The wine containing little or no unfermented sugar is known as---------------

(a) Sherry (b) Dry wine (c) Sake (d) Red wine

54. Which of the following is another name for an intact virus particle?

(a) Nucleocapsids (b) Capsides (c) Virions (d) None of the above

55. Phycology is the study of -------------

(a) Fungi (b) Algae (c) Protozoa (d) Nucleic acid

56. ------------is a symbiosis in which two or more organisms living together benefit each other.

(a) Parasitism (b) Commensalism (c) Mutualism (d) None of the above

57. Pink rots on eggs are produced due to the growth of ---------------

(a) Clostridium (b) Pseudomonas (c) Streptococcus (d) Penicillium

58. ----------------- is the chief source for spoilage of poultry.

(a) Algae (b) Actinomycetes (c) Fungi (d) Bacteria

59. Slime layer production in meat is due to-------------

(a) *Streptococcus* (b) *Moraxella*

(c) *Clostridium* (d) *Leuconostoc mesenteriod*

60. In the spoilage of meat, changes in fat occurs due to--------------

(a) Achromobacter (b) Luminous bacteria (c) Actinomycetes (d) Proteolytic bacteria

61. Green blue colour of meat is due to-----------------

(a) *Serratia morcescens* (b) *Pseudomonas syncynea*

(c) *Chromobacterium lividum* (d) *Micrococcus*

62. Oyster mushroom is the common name of which species of mushroom?

(a) *Pleurotus ostreatus* (b) *Agaricus bisporus*

(c) *Leutinule edodes* (d) None of the above

63. Growth of mushroom spawn is called as---------------

(a) Spawn running (b) Spawning (c) Cropping (d) Casing

64. In biogas which main gas used is?

(a) O_2 (b) Methane (c) CO_2 (d) N_2

65. Which of the following are the characteristics of Entomopathogenic virus?

(a) Can be easily produced

(b) Effective pathogen

(c) Active against economically important insects

(d) Low in virulence, unstable and difficult to produce

66. The type of pathogenecity in which enteropathogen is generally administrated once and animal is observed for short period is known as-----------------

(a) Autotoxicity (b) Sub autotoxicity (c) Carcinogenicity (d) Primary irritation

67. The time in minutes at a specific temperature needed to kill a population of cells is the-------
    (a) Decimal reduction time (b) Thermal death point
    (c) F value (d) D value
68. Glutamic acid production from non carbohydrate material can be done with the help of which microorganism?
    (a) *Corneybacterium glutamicum* (b) *Brevibacterium flaum*
    (c) *Candida guilliermondii* (d) *E. Coli*
69. L-leucine is produced by the action of ---------------
    (a) *Serratia marescens* (b) *Bacillus substillus*
    (c) *Corneybacterium glycinophilum* (d) *Candida utilis*
70. ---------------is the important vitamin used for the production of baker's yeast.
    (a) Riboflavin (b) Thiamine (c) Biotin (d) None of the above
71. Alcohol content of table wine is---------------%.
    (a) 20-25 (b) 8-15 (c) 15-20 (d) 3-8
72. Which microorganism is used for the production of sherry?
    (a) *Saccharomyces bayanus* (b) *Sacchromyces cerevasae*
    (c) *Sacchromyces ellipsoideus* (d) All of the above
73. Very pale coloured beer which is fully attenuated with high level of hops is called as---------
    (a) Dortmunder (b) Munchener (c) Ales (d) Pilsner
74. Which of the following is not top fermented beer?
    (a) Munich (b) Pale (c) Mild (d) Stout
75. Which of the following steroid is progesterone?
    (a) *Rhizopus arrhizus* (b) *Fusarium solani*
    (c) *St.albus* (d) *Alkaligens faecalis*
76. Which of the following disinfectants acts by disrupting microbial membranes?
    (a) Cationic detergents (b) Halogens
    (c) Heavy metals (d) Aldehydes
77. Cheddar cheese is produced due to the action of which microorganism?
    (a) *Saccharomyces lactis* (b) *Lactobacillus bulgaricus*
    (c) *Streptococuss thermophilus* (d) *Penicillium roqueforti*
78. Which of the following prevent the germination of *Clostridium botulinum?*
    (a) Sorbic acid (b) Benzoic acid (c) Nitrate and nitrites (d) SO_2
79. Which of the following genus of bacteria oxidizes ethyl alcohol to acetic acid?
    (a) *Acetobactor* (b) *Arthrobactor* (c) *Bacillus* (d) *Enterobactor*
80. *Zygosaccharomyces* is the example of which type of yeast?
    (a) Top yeast (b) Bottom yeast (c) Osmophilic yeast (d) Halophilic yeast

81. Which of the following is true yeast?
(a) Candida truste (b) Rhodtonua (c) Brettanomyces (d) Pichia

82. Single celled spores formed at the end of special hyphae are called as---------------
(a) Sporangiospores (b) Canidiospores (c) Arthrospores (d) Chlamydospores

83. Process of fusion of two cells and their protoplast is known as-----------------
(a) Oogamy (b) Kayogamy (c) Plasmogamy (d) Meiosis

84. Fusion of somatic or vegetative cell in mold is known as---------------
(a) Spermatization (b) Gametic copulation
(c) Gametangial copulation (d) Somatic copulation

85. Transfer of genes from one cell to another by a bacterio phase is known as-------------
(a) Conjugation (b) Transduction (c) Transformation (d) None of the above

86. The organism which uses organic compounds as electron donor is known as--------------
(a) Organotrophs (b) Liphotrophs (c) Chemolithotrophs (d) None of the above

87. Who introduced the antiseptic method for wound infection?
(a) Robert Koch (b) Schawann (c) Joseph Lister (d) Jontyndall

88. The initial growth phase, during which cell number remains relatively constant prior to rapid growth is known as--------------
(a) Log (b) Lag
(c) Stationary (d) Phase of positive acceleration

89. *Saccharomyces cerevisiae* is being grown in a chemostat converts glucose to biomass, ethanol, glycerol and carbon dioxide. At steady state, the concentration of glucose, biomass, ethanol and glycerol will---------------
(a) Decrease with time (b) Increase with time
(c) Be constant (d) Change randomly with time

90. The cell without cell wall is called as-------------
(a) Cytoplast (b) Organoplast (c) Protoplast (d) None of the above

91. Who is called as father of Biotechnology?
(a) D.R. John (b) Karl Ereky (c) J.D Neit (d) Kreit Thomson

92. The biotechnology which is applied to the agricultural processes is called as ------------
(a) Blue biotechnology (b) Red biotechnology
(c) White biotechnology (d) Green biotechnology

93. Who is the father of tissue culture?
(a) Roff Harrison (b) Moud Mentel (c) Broed Deake (d) Alexander Ruzvelt

94. The lowest biomass yield in a culture of *Escherichia coli* will be in------------
(a) An aerated batch culture containing a initial high concentration of glucose
(b) An aerated batch reactor containing an initial low concentration of glucose

(c) An aerated fed-batch reactor having a low glucose concentration
(d) An aerated continuous reactor having a low glucose concentration

95. Which of the following is physical sanitizing agent?
(a) Alcohol (b) Acid (c) Hot water (d) None of the above

96. Sodium hydroxide is---------------
(a) Acid cleaner (b) Alkaline cleaner (c) Both a and b (d) Neutral

97. Which disinfectant would be most effective against *Staphylococcus* found in a blood spill?
(a) Phenol (b) Cetylpyridinium chloride
(c) Hexachlorophene (d) None of the above

98. ------------- is known as bread mold.
(a) *Penicillium notatum* (b) *Aspergillus niger*
(c) *Rhizopus stolonifer* (d) *Monilla sitophilla*

99. Softness of pickle is due to--------------
(a) *Penicillium* (b) *Bacillus* (c) *Lactobacillus* (d) *Pseudomonas*

100. Pectolysis of pickles is due to---------------
(a) *Aeromonas* (b) *Achromobacter* (c) Both a and b (d) None of the above

101. What is ale?
(a) Fermented corn (b) Type of beer (c) Fermented carrot (d) None of the above

102. Which of the following microorganisms is used in beer making?
(a) Yeast (b) Bacteria (c) Mold (d) All of the above

103. Hopes are used in manufacturing of--------------
(a) Wine (b) Beer (c) Brandy (d) Whiskey

104. Antibiotics are the ------------- metabolic products of the organism.
(a) Primary (b) Secondary (c) Tertiary (d) Organic

105. Penicillin interferes with---------------
(a) Cell membrane (b) Peptidoglyoan synthesis
(c) Translocation action (d) Nucleotide synthesis

106. ------------ is the process of transfer of genetic material between bacterial cells by direct cell to cell contact.
(a) Transmation (b) Transduction (c) Conjugation (d) Transfection

107. Fungi are -------------
(a) Heterotrophic (b) Autotrophic (c) Chemotrophic (d) Chemoautotrophic

108. Organism that can tolerate high sugar concentration are called as-------------
(a) Halotolerant (b) Halophlic (c) Xerotolerant (d) Xerophilic

109. Study of fungi is called---------------
(a) Phycology (b) Zoology (c) Virology (d) Mycology

110. Which of the following is most effective for sterilizing mattresses and plastic petri plates?
(a) Chlorine (b) Ethylene oxide (c) Glutaraldehyde (d) Autoclave

111. Which disinfectant would be most effective against *Bacillus cereus*?
(a) Phenol (b) Cetylpyridinium chloride
(c) Hexachlorophene (d) Merthiolate

112. The pink or reddish liquid that comes from meat on thawing is called----------------
(a) Drip (b) Leakage (c) Marbeling (d) Oozing

113. -------------species is responsible for slime in sugarcane juice.
(a) Bacillus (b) Proteus (c) Leuconostoc (d) Streptococi

114. ------------ is the mycotoxin produced by *Penicillium patulum*.
(a) Aflatoxin (b) Patulin (c) Ochratoxin (d) Luteoskyrin

115. Most active sanitizer of the chlorine compounds is-------------
(a) Chlorine dioxide (b) Inorganic chloramines
(c) Organic chloramine (d) Hypochlorites

116. Most popular form of iodine compound used as sanitizer is--------------
(a) Iodophor (b) Aqueous
(c) Alcohol–iodine solution (d) Free elemental iodine

117. Scouting compounds are also known as--------------
(a) Chemical abrasives (b) Physical abrasives
(c) Surface active agents (d) Cleaning compounds

118. ----------------- is the medium which contains fermentable sugars and hops extract.
(a) Must (b) Fruit juice (c) Wort (d) Molasses

119. ------------------may cause a yellow colour in the cream layer of milk.
(a) *Pseudomonas synxantha* (b) *Pseudomonas syncyanea*
(c) *Pseudomonas aeruginosa* (d) *Serratia marcescens*

120. An autoclave makes use of ------------ for sterilization.
(a) Moist heat (b) Dry heat (c) Radiation (d) Filtration

121. -------------- staining is widely used differential staining procedure in bacteriology.
(a) Acid fast (b) Simple (c) Gram (d) None of the above

122. Ozone has a strong microbicidal action against---------
(a) Bacteria (b) Fungi
(c) Parasites and viruses (d) All of the above

123. Ozone can be used in food industry as-----------
(a) Disinfection (b) Deodorization (c) Decolourization (d) All of the above

124. Which microorganism is used in production of citric acid?
(a) *A. niger* (b) *B. subtilis* (c) *cl. terani* (d) *E. coli*

125. Lactic acid is produced by which microorganism?
(a) *Lactobacillus bulgaricus* (b) *A.niger*
(c) *S. cerevisiae* (d) *E. coli*

126. ------------ is the female flower added in the beer for bitter taste and aroma.
(a) Wort (b) Autibiotic (c) Hops (d) None of the above

127. ------------------is produces inside the cell and secreted outside in the medium.
(a) Endoezyne (b) Exoenzyme (c) Both a and b (d) None of the above

128. Mycotoxins are produced by ---------------
(a) Fungi (b) b. Algae (c) Virus (d) Bacteria

129. --------------- contains only one specific type of microorganism.
(a) Mixed culture (b) Pure culture (c) Culture (d) None of the above

130. Which microorganism is used in *temph* preparation?
(a) *Rhizopus stolonifer* (b) *B. subtilis*
(c) *A. niger* (d) *S. cerevisiae*

131. ------------ is produced from fermented apple juice.
(a) Cider (b) Temph (c) Koji (d) Kefir

132. Agar is used in microbial media as ---------------
(a) Solidifying agent (b) Binding agent (c) Nitrogen source (d) Carbon source

133. ------------ is used as a precursor in the production of antibiotics.
(a) Phenyl acetic acid (b) Phenol
(c) Formalin (d) None of the above

134. Penicillin is made by employing --------------- fermentation.
(a) Continuous (b) Aerobic (c) Anaerobic (d) None of the above

135. Production of alcohol by fermentation of molasses is an -------------process.
(a) Anaerobic (b) Aerobic (c) Endothermic (d) All of the above

136. Which of the following statements is/ are correct?
(a) Generation time is the time required by the cell to double
(b) Generation time isn't same for all microbes
(c) Generation time isn't same for a microbe under all conditions
(d) All of the above

137. Iodophors are generally more effective than chlorine even with------------
(a) Organic matter (b) Water hardness (c) Both a and b (d) None of the above

138. Ozone can be used in food industry for ----------
(a) Sanitation of food plant equipment
(b) Treatment and lowering biological oxygen demand
(c) Chemical oxygen demand of food plant waste
(d) All of the above

139. Ozone is a strong ------------

(a) Oxidant (b) Potent disinfecting agent

(c) Sanitizer (d) All of the above

140. Which of the following statements with respect to growth of bacteria is/ are correct?

(a) Growth rate is reciprocal to generation time

(b) Generation time of *B. Subtilis* is 8 min. at 37^0C in broth

(c) Both a and b

(d) None of the above

141. Which of the following statements is/ are not correct for logarithmic phase of bacterial growth curve?

(a) Growth rate is not constant (b) Number of cells remains constant

(c) It is the phase of cell enlargement (d) All of the above

142. The dye used in reduction test is/ are-----------------

(a) Resazurin (b) Methylene blue (c) Both a and b (d) None of the above

143. Total plate count method is being expressed in----------------

(a) No. of cells/ ml (b) CFU/ ml (c) Optical density (d) None of the above

144. Most heat resistant spores are of----------------

(a) *Staphylococcus aureus* (b) *Clostidium sporogenes*

(c) *Clostidium botulinum* (d) *Bacillus stearothermophilus*

145. Iodine is highly temperature-dependent and vaporizes at--------------°F.

(a) 120 (b) 180 (c) 160 (d) 140

146. ------------------- are most common causes of spoilage of fruits and vegetables.

(a) Molds (b) Bacteria (c) Yeasts (d) All of the above

147. Sliminess or souring is due to--------------

(a) Mold (b) Osmophilic yeast

(c) Saprophytic bacteria (d) Saccharophilic mold

148. In jams and jellies, which of the following is most likely to be responsible for the spoilage?

(a) Bacteria only (b) Yeasts and molds

(c) Molds only (d) Yeasts, molds and bacteria

149. Undesirable odour in sauerkraut is due to the action of which of the following microbe?

(a) *Lactococcus brevis* (b) *Pediococcus cerevisiae*

(c) *Leuconostoc mesenteroides* (d) All of the above

150. Actinomycetes are responsible for --------------- in meat.

(a) Souring (b) Taints (c) Stamping ink (d) Earthy flavour

151. Putrefaction is caused by--------------

(a) *Alkaligens spp.* (b) *Geotrichum spp.* (c) *Clostridium spp.* (d) *Thamnidium spp.*

152. Superficial fungal spoilage in egg is at----------------
(a) High temperature (b) High humidity
(c) Low temperature (d) Low humidity

153. Red rot of egg is due to --------------
(a) *Serratia spp.* (b) *Cladosporium spp.* (c) *Pseudomonas spp.* (d) *Sporotrichum spp.*

154. The milk is pasteurized at 62.8^{0}C for 30 min to eliminate-------------
(a) *Mycobacterium tuberculosis* (b) *Listeria monocytogens*
(c) *Coxiella burnetti* (d) *Salmonella spp.*

155. Iodophor is mixtures of-----------
(a) Iodine and surfactant as a carrier (b) Iodine and cleaning agent as a carrier
(c) Iodine and phenol as a carrier (d) All of the above

156. Generally recommended usage for iodophors is ---------
(a) 12.5 to 25ppm for 1 min (b) 22.5 to 25ppm for 1 min
(c) 32.5 to 25ppm for 1 min (d) 42.5 to 25ppm for 1 min

157. Sweet curdling in milk is caused by ---------------
(a) *Alcaligens* viscolactis (b) *Bacillus cereus*
(c) *Enterobacter aerogens* (d) *Pseudomonas fluorescens*

158. Ropiness of the milk which is observed at the top is caused by--------------
(a) *Alcaligens viscolactis* (b) *Bacillus cereus*
(c) *Enterobacter aerogens* (d) *Pseudomonas fluorescens*

159. Ropiness of the milk which is observed throughout the milk is caused by------------
(a) *Alcaligens* viscolactis (b) *B. cereus*
(c) *Enterobacter aerogens* (d) *Pseudomonas fluorescens*

160. Which type of spoilage do you expect in canned evaporated milk?
(a) Curd formation (b) Bitter flavour (c) Bulged can (d) All of the above

161. Rusty spot in cheese is due to---------------
(a) *Lactobacillus bulgaricus* (b) *Bacillus subtilis*
(c) *Lactobacillus plantarum* (d) *Leuconostoc mesenteroides*

162. Red milk is caused by-------------
(a) *Pseudomonas syncyanea* (b) *Pseudomonas synxantha*
(c) *Serratia marcescens* (d) *Pseudomonas putrefaciens*

163. Brown milk is caused by------------
(a) *Pseudomonas syncyanea* (b) *Pseudomonas synxantha*
(c) *Serratia marcescens* (d) *Pseudomonas putrefaciens*

164. Action of food acids on the can result in a --------------
(a) Hard swell (b) Soft swell (c) Hydrogen swell (d) Nitrogen swell

165. *Clostridium botulinum* is --------------- type of microorganism.
(a) Mesophilic non- spore forming (b) Mesophilic spore forming
(c) Thermophilic non- spore forming (d) Thermophilic spore forming

166. Ropiness in wine is caused by--------------
(a) *Lactococcus brevis* (b) *Pseudomonas salineria*
(c) *Leuconostoc mesenteroides* (d) All of the above

167. ----------------is less affected by water hardness when compared to other sanitizers.
(a) Chlorine
(b) Quaternary ammonium compounds
(c) Concern is based upon the involvement of chlorine in the formation of potentially
(d) Carcinogenic trihalomethanes

168. The alcoholic beverage made by distillation of the fermented juice of cereal grains is known as -----------------
(a) Whiskey (b) Brandy (c) Rum (d) Vodka

169. Antibiotics are produced during ------------ phase of growth of the microorganisms.
(a) Lag (b) Stationary (c) Logarithmic (d) Death

170. In the fermenter, aeration is achieved by using-------------
(a) Impeller (b) Baffles (c) Sparger (d) Biosensors

171. The agent that helps in penetration of the stain into the cell part to be stained is----------
(a) Fixative (b) Accentuator (c) Mordant (d) Chromogen

172. An example of differential staining is-------------
(a) Monochrome staining (b) Negative staining
(c) Gram staining (d) Endospore staining

173. -------------- are the cell forms that can be prepared from gram positive bacteria by treating the cells with an enzyme lysozyme.
(a) Protoplasts (b) L- forms (c) Spheroplasts (d) None of the above

174. Most effective chlorine compounds include-----------
(a) Liquid chlorine, hypochlorites (b) Inorganic chloramines, and organic chloramines
(c) Hypochlorous acid (d) All of the above

175. The activity of chlorine is dramatically affected by------------
(a) pH (b) Temperature (c) Organic load (d) All of the above

176. *Monotrichous flagella* are------------
(a) Cluster of flagella at one end (b) Single flagellum at one end
(c) Flagella present laterally (d) Cluster of flagella at both the ends

177. Which one of the following is not a mechanism of gene transfer between the bacterial cells?
(a) Transformation (b) Transduction (c) Conjugation (d) Mutation

178. Virulent phages is/ are--------------
(a) Phages following lytic cycle exclusively (b) Phages following lysogenic cycle exclusively
(c) Phages following both the cycles (d) None of the above

179. -------------- is the shrinkage of cell contents as a result of withdrawal of water by osmosis.
(a) Plasmolysis (b) Osmolysis (c) Autolysis (d) All of the above

180. The ideal chemical sanitizer should---------
(a) Be approved for food contact surface application
(b) Have a wide range or scope of activity
(c) Destroy microorganisms rapidly
(d) All of the above

181. Linear inhibition is sometimes called as---------------
(a) Complete inhibition (b) Incomplete inhibition
(c) Partial inhibition (d) Mixed inhibition

182. Ang- kak is produced commercially by using---------------
(a) *Monascus purpureus* (b) *E. coli*
(a) *A. niger* (d) *Penicillium notatum*

183. Which one of the following is distilled wine?
(a) Brandy (b) Gin (c) Whiskey (d) Vodka

184. Which one of the following is not an example of mycotoxin?
(a) Aflatoxin (b) Ochratoxin (c) Patulin (d) Botulin

185. ---------------is a type of wine which does not retain any of the CO_2 produced.
(a) Sweet wine (b) Table wine (c) Still wine (d) Fortified wine

186. -------------- means alternate forms of genes.
(a) Alleles (b) Clone (c) Both a and b (d) None of the above

187. -------------- is a group of Gram negative, prokaryotic, obligate intracellular parasites of arthropods.
(a) Mycoplasma (b) Actinomycetes (c) Rickettsia (d) All of the above

188. Karyogamy is -------------
(a) Union of protoplast (b) Union of haploid nucleus
(c) Union of egg and sperm (d) None of the above

189. Rum is manufactured from-------------
(a) Grape juice (b) Rice (c) Sugarcane molasses (d) Carrot juice

190. During manufacture of wine, the crushed grapes used as raw material is called-------------
(a) Wort (b) Hops (c) Ale (d) Must

191. Red wine is obtained by fermentation of ---------------
(a) Banana juice (b) Grapes with skin
(c) Grapes by red pigment producing bacteria (d) Red roses

192. Ale is manufactured by fermentation using---------------
(a) Top yeasts (b) Bottom yeasts
(c) Thermophilic molds (d) Capsulated bacteria

193. Brewing is---------------
(a) Cultivation of grapes (b) Preservation of meat by fermentation
(c) Manufacture of beer (d) None of the above

194. Hops are effective against-----------------
(a) Gram positive bacteria
(b) Gram negative bacteria
(c) Gram positive as well as gram negative bacteria
(d) Not effective against bacteria

195. Dry wines are free from --------------
(a) Moisture (b) Toxic substances
(c) Unfermented sugars (d) All of the above

196. Which one of the following is an example of bacterial food- borne illness?
(a) Polio (b) Botulism (c) Hepatitis (d) All of the above

197. Which of the following act as a cloning vector?
(a) Plasmids (b) Cosmids (c) Phag (d) All of the above

198. Sparkling wine contains--------------
(a) Carbon dioxide (b) Sodium benzoate (c) Fruit skin (d) All of the above

199. Siphoning off of the fermented wine to separate it from the solid deposits is known as-------
(a) Racking (b) Ageing (c) Maturation (d) None of the above

200. Maturation of wine is also known as------------
(a) Fermentation (b) Ageing (c) Fining (d) All of the above

201. Clear, sparkling wine is called as--------------
(a) Champagne (b) Perry (c) Sherry (d) None of the above

202. Ropiness in bread is caused by--------------
(a) *Bacillus subtilis* (b) *Serratia marcescens*
(c) *Geotrichum auranticum* (d) None of the above

203. Flat sour spoilage of canned food is caused due to---------------
(a) *Pseudomonas* (b) *Bacillus* (c) *Clostridium* (d) *Serratia*

204. TA spoilage of canned food is caused by ---------------- bacteria
(a) Thermophillic, aerobic (b) Thermophillic, anaerobic
(c) Thermoduric, aerobic (d) Thermodeuric ,anaerobic

205. Which variety of *Sacchromyces cerevisae* is used in beer manufacture?
(a) *Mesentericus* (b) *Ellipsoideus* (c) *Vermiformis* (d) *Carlsbergensis*

206. Rum is---------------
(a) Distilled liquor (b) Undistilled liquor
(c) Fortified wine (d) Byproduct of brewing industry

207. The thread like worms found in vinegar are called----------------
(a) Vinegar eels (b) Vinegar louse (c) Vinegar mites (d) Vinegar flies

208. The mould which causes spoilage of canned fruit is--------------
(a) *Penicillium sp.* (b) *Aspergillus sp.* (c) *Mucor sp.* (d) *Byssochlamys fulva*

209. Which type of fermentation occurs in preservation of fruits and vegetables by pickling?
(a) Acetic acid fermentation (b) Lactic acid fermentation
(c) Alcoholic fermentation (d) All of the above

210. In the process of wine preparation which type of fermentation takes place?
(a) Acetic acid fermentation (b) Lactic acid fermentation
(c) Alcoholic fermentation (d) All of the above

211. In canned foods sulphide spoilage is caused by----------------
(a) *Bacillus coagulans* (b) *Bacillus sterothermophilus*
(c) *Clostridium nigrificans* (d) *Clostridium thermosaccharolyticum*

212. Which of the following is a fermented beverage?
(a) Nectar (b) Cordial (c) Crush (d) Wine

213. The wines containing added alcohol in the form of brandy is known as--------------
(a) Dry wines (b) Sweet wines (c) Sparkling wines (d) Fortified wine

214. The major spp. responsible for rotting of citrus fruit is--------------
(a) Alternaria (b) Fusarium (c) Penicillum (d) None of the above

215. The culture used for production of sweet *dahi* is----------------
(a) *S. lactis* (b) *S. diacetilactis* (c) *S. cremoris* (d) All of the above

216. Spoilage in food products occur due to-----------------
(a) Enzymes in food (b) Microbes in food (c) Both a and b (d) None of the above

217. Surface slime in meat is caused due to---------------
(a) *Pseudomonas* (b) *Alcaligenes* (c) *Leuconostoc* (d) All of the above

218. Whiskers in meat is caused due to-----------------
(a) *Thamnidium* (b) *Mucor* (c) *Rhizopus* (d) All of the above

219. Souring in food is caused due to------------------
(a) Production of organic acids (b) Production of inorganic acids
(c) Production ascorbic acid (d) All of the above

220. Proteolysis of milk is caused due to------------
(a) *Micrococcus* (b) *Alcaligenes* (c) *Proteus* (d) All of the above

221. TA spoilage in cans is caused due to---------------
(a) *Cl. perfringens* (b) *Cl. thermosaccharolyticum*
(c) *Cl. botulinum* (d) Both a and b

222. Gray mold rot in grapes is caused due to---------------
(a) *B. cinera* (b) *R. stolonifer* (c) *A. niger* (d) None of the above

223. The infectious disease which spreads from one continent to other is termed as-------------
(a) Pandemic (b) Sporadic (c) Epidemic (d) None of the above

224. Effectiveness of sanitizers is reduced by-----------------
(a) Organic residues on surface of equipments
(b) Improper rinsing of the equipments
(c) Improper handling of the equipments
(d) Both a and b

225. The effectiveness of thermal sanitizing is dependant upon ---------------
(a) Initial contamination load (b) Humidity, pH
(c) Temperature, and time (d) All of the above

226. CIP stands for---------------
(a) Cleaning in place (b) Cleaning improvement project
(c) Cleaning in processing (d) None of the above

227. Epidemic starts from-------------
(a) Star hotels (b) Urban areas (c) Slum areas (d) Normally from animals

228. Alum is used in purification of water---------------
(a) To kill pathogens (b) To kill non pathogens
(c) To remove organic matter (d) None of the above

229. Which one of the following is not oxidative yeast?
(a) *Pichia* (b) *Kloekera* (c) *Saccharomyces* (d) *Candida*

230. Iodine sanitizers causes death of pathogens, due to--------------
(a) Destruction of mitochondria (b) Damage to DNA
(c) Cell wall damage (d) None of the above

231. The chief spoilage organisms on smoked fish are--------------
(a) Molds (b) Bacteria (c) Both a and b (d) Fungi

232. Hypochlorite's are more effective at----------------
(a) Acidic pH (b) Neutral pH (c) Alkaline pH (d) None of the above

233. The non-volatile compound in oak wood barrel is--------------
(a) Phenolic acid (b) Phenol aldehyde (c) Vanillin (d) Eugenol

234. Quaternary ammonium compounds are in general more effective against-------------
(a) Gram +ve (b) Gram –ve (c) Spore formers (d) None spore formers

235. The cleaning agent should have property/properties like------------

(a) Emulsification (b) Wettability (c) Rinsibility (d) All of the above

236. Disinfection of sewage water is a---------------

(a) Primary treatment (b) Secondary treatment

(c) Tertiary treatment (d) None of the above

237. Ergotin mycotoxin is produced by--------------

(a) *Penicillium notatum* (b) *Aspergillus niger*

(c) *Claviceps purpurea* (d) *Rhizopus nigricans*

238. The first outbreak of bird flu was recorded in-------------

(a) 1997 (b) 1998 (c) 1999 (d) 2000

239. Bird flu is caused due to--------------

(a) H_5N_1 (b) $H_{10}N_{11}$ (c) H_1N_1 (d) None of the above

240. Round worm parasites belongs to------------

(a) Trematodes (b) Cestodes (c) Nematodes (d) None of the above

241. Which of the following does not have antimicrobial activity?

(a) Chlorophyll (b) Organic acids (c) Spice extracts (d) Hydrogen peroxide

242. In preparation of curd from milk the fermentation is terminated due to------------

(a) Complete oxidation of milk sugar (b) Death of fermenting organisms

(c) Fall in pH which inhibit microbial activity (d) None of the above

243. Which of the following aflatoxins found in food is the most potent toxin?

(a) B_1 (b) B_2 (c) G_1 (d) G_2

244. The conditions of pasteurization of milk are worked out based on destruction of------------

(a) *Mycobacterium tuberculosis* (b) *Salmonella typhimurium*

(c) *Clostridium botulinum* (d) *Staphylococcus aureus*

245. Groundnut is prone to incidences of contamination with--------------

(a) *Aspergillus flavus* (b) *Aspergillus paraciticus*

(c) *Aspergillus niger* (d) Both a and b

246. Sour flavour in milk during storage is caused by-------------

(a) *Alkaligenes viscolactis* (b) *Streptococcus lactis*

(c) *Pseudomonas suncyanea* (d) *Bacillus subtilis*

247. Flat sour in cans is caused due to------------

(a) *Pseudomonas spp.* (b) *Clostridium spp.*

(c) *Bacillus spp.* (d) None of the above

248. The disease which spreads from one continent to other is termed as------------

(a) Pandemic (b) Sporadic (c) Epidemic (d) None of the above

249. Effectiveness of sanitizers is reduced by----------------
(a) Organic residues on surface of equipments
(b) Improper rinsing of the equipments
(c) Improper handling of the equipments
(d) Both a and b

250. Which one of the dairy product is prepared with pure culture of *L. delbrueckii*?
(a) Indian buttermilk (b) Bulgarian buttermilk
(c) Cultured buttermilk (d) None of the above

251. The microorganisms which requires very less moisture for growth are called-------------
(a) Mesophilic (b) Xerophilic (c) Halophilic (d) Thermophilic

252. The optimum temperature for growing psychrophillic bacteria is-------------°C.
(a) 0 (b) 15 (c) 60 (d) 45

253. Amoebic dysentery is caused due to---------------
(a) *Ascaris lumbricoides* (b) *Entamoeba histolytica*
(c) *Ancylostoma duodenale* (d) Both a and c

254. Wine flower is--------------
(a) Type of wine fortified with flower extract
(b) Type of defect found in wine
(c) Type of wine supplement prepared from flower extract
(d) None of the above

255. Trickling filters are used--------------
(a) Clarification of syrups (b) Clarification of fruit juices
(c) Clarification of sewage (d) None of the above

256. In bird flu causing virus H and N stands for-------------
(a) Haemaglutinin and Neuraminidase (b) Haematoglutinin and Nitrous oxide
(c) Haemoglobin and nucleic acid (d) None of the above

257. Which of the following microorganisms cannot tolerate oxygen?
(a) *Clostridium botulinum* (b) *Staphylococcus aureus*
(c) *Penicillium roquefortii* (d) *E. coli*

258. --------------consists of a live microbial food additives together with prebiotic oligosaccharides.
(a) Antibiotics (b) Prebiotics (c) Synbiotics (d) Both a and b

259. Barley is ideally suited to brewing due to-------------
(a) High complement of enzymes (b) Contain good amount of proteins
(c) Husk helps in filtration (d) All of the above

260. Which of the following type of barley is used in brewing?
(a) Two row and six row (b) Two row and four row
(c) Six row and four row (d) All of the above

261. European brewers prefer which type of barley for brewing?
(a) Two row (b) Four row (c) Six row (d) All of the above

262. American brewers prefer which type of barley for brewing?
(a) Two row (b) Four row (c) Six row (d) All of the above

263. Which type of barley has high levels of starch degrading enzymes?
(a) Two row (b) Six row (c) Four row (d) All of the above

264. Six row barley has---------------
(a) More proteins (b) Less starch (c) Thicker husk (d) All of the above

265. In beer, husk of the malt contributes for---------------
(a) Higher polyphenols (b) Haze in beer
(c) Imparts an astringent taste (d) All of the above

266. Which type of barley has lower enzyme content?
(a) Two row (b) Four row (c) Six row (d) All of the above

267. Which type of barley has higher levels of starch?
(a) Six row (b) Four row (c) Two row (d) None of the above

268. Malting helps in generating---------------
(a) Starch to simple sugars (b) Reducing protein complexes
(c) Development of enzymes (d) All of the above

269. During steeping, barley is turned several times------------
(a) To avoid microbial growth (b) To increase oxygen uptake by barley
(c) To slow down rate of water uptake (d) None of the above

270. Mechanical designs for pneumatic malting includes--------------
(a) Compartment and continuous (b) Tower and circular
(c) Flexi malting (d) All of the above

271. Malt kilning at high temperature improves------------
(a) Colour and flavour of malt (b) Enzyme activity of malt
(c) Fatty oil content of malt (d) All of the above

272. ASBC Stands for------------
(a) Asian Society of Brewing Chemists (b) American Society of Brewing Chemists
(c) Australian Society of Brewing Chemists (d) None of the above

273. Barley flavour is influenced by-----------
(a) Carbon dioxide (b) Adjuncts (c) Malt (d) Both b and c

274. Adjuncts used in brewing results in--------------
(a) Enhanced stability of beer (b) Superior chill proof quality
(c) Both a and b (d) None of the above

275. During wort boiling what happens to proteins?
(a) Coagulation (b) Hydrolysis (c) Solubilization (d) All of the above

276. During boiling of wort, colour become darker due to----------------
(a) Formation of melanoidins (b) Oxidation of polyphenols
(c) Caramelization of sugars (d) All of the above

277. Major flavour contribution by hops is due to--------------
(a) Iso-alpha acids (b) Dimethyl sulphides (c) Aldehydes (d) All of the above

278. Some breweries, formation of hot break is enhanced by addition of--------------
(a) Calcium sulphate (b) Lactic acid
(c) Tannic acid (d) None of the above

279. For bottom fermented beer, wort is cooled to------------°C
(a) 2 to 5 (b) 5 to 15 (c) 15 to 25 (d) 25 to 30

280. Pasteurization is used to destroy ---------------
(a) Antibiotics (b) Bacteria (c) Somatic cells (d) None of the above

281. Microorganism involved in yoghurt formation are--------------
(a) *Vibrio parahaemolyticus* (b) *Streptococcus*
(c) *Lactobacillus bulgaricus* (d) Both b and c

282. Which of the following organism contaminate meat, poultry and fish products?
(a) *Salmonella* (b) *Pseudomonas* (c) *Staphyllococci* (d) *Lactobacilus*

283. In TA spoilage, gas formed is---------
(a) Hydrogen sulphide (b) Carbon monoxide
(c) Hydrogen (d) Nirogen

284. Chemical sanitizers containing---------- are most widely used for sanitizing milking equipments.
(a) Bromine (b) Saline (c) Chlorine (d) Iodine

285. Which of the following is a bacterial intoxication?
(a) Salmonelosis (b) Botulism (c) Shigellosis (d) Gas gangrene

286. AID's is ---------------- disease.
(a) Bacterial (b) Viral (c) Fungal (d) Both a and b

287. Aflatoxin M is found in ---------------
(a) Groundnut (b) Wheat (c) Milk (d) Soybean

288. Which gas is evolved during fermentation?
(a) Oxygen (b) CO_2 (c) Methane (d) All of the above

289. Insoluble complexes formed due to microbial colonies or protein–tannin complexes in wine, beer or juices is called as---------------
(a) Haze (b) Foam (c) Blaze (d) SPM

290. Which of the following is microbial gum?
(a) Algin (b) Xanthan (c) Gum Arabic (d) Agar

291. Which of the following microorganisms cannot tolerate oxygen?
(a) *Clostridium botulinum* (b) *Staphylococcus aureus*
(c) *Penicillium roquefortii* (d) *E. coli*

292. Fluidized bed bioreactors provide higher mass transfer rates than packed bed bioreactors because--------------
(a) Mixing is higher in fluidized bed bioreactors
(b) Particles move with the fluid in a fluidized bed bioreactor
(c) Immobilized particles are smaller in the fluidized bed bioreactors
(d) All of the above

293. Which microorganism is commonly associated with fecal contamination?
(a) *Clostridium botulinum* (b) *Campylobacter jejuni*
(c) *Bacillus cereus* (d) *Trichinella spiralis*

294. *Staphylococcus aureus* is --------------
(a) Gram –ve cocci (b) Gram +ve cocci (c) Gram +ve rod (d) Gram –ve rod

295. Gastroenteritis is caused due to---------------
(a) *Vibrio cholerae* (b) *Vibrio parahaemolyticus*
(c) Both a and b (d) None of the above

296. No fumigants shall be used directly on article of food except registered and approved under Insecticide Act -----------
(a) 1966 (b) 1967 (c) 1968 (d) 1969

297. If biomass yields are constant, then the biomass productivity of a culture grown in continuous reactor will------------
(a) Increase with dilution rate until washout
(b) Always decrease with dilution rate
(c) Remain constant irrespective of the dilution rate
(d) Decrease with dilution rate until washout

298. Plasmids can provide bonus genetic information for -----------
(a) Antibiotic resistance (b) Virulence factors
(c) Promote conjugation (d) All of the above

299. Virus particles ----------------
(a) Are not made of cells (b) Cannot reproduce on their own
(c) Do not transform energy (d) d. All of the above

300. Viroids are uniquely characterized by ---------------
(a) The presence of a capsid (b) Can reproduce on their own
(c) The absence of a capsid (d) Non of the above

301. Bacteria causing spoilage of fish are----------------

(a) *Botrytis* and *Mucor* (b) *Pseudomonas* and *Flavobacterium*
(c) *Aerobacter* and *Lactobacillus* (d) None of above

302. *Schizosaccharomyces pombe* reproduces asexually by-----------

(a) Budding (b) Fission and arthrospores
(c) Budding and ascospores (d) Fission

303. The white cottony growth of molds is called-----------

(a) Spores (b) Mycelium (c) Hypha (d) None of the above

304. Round worm parasites belongs to----------------

(a) Trematodes (b) Cestodes (c) Nematodes (d) None of the above

305. Which of the following is bactericidal (lethal to bacteria)?

(a) Membrane filtration (b) Ionizing radiation
(c) Deep freezing (d) All of the above

306. Which of the foods listed would be most likely to spoil as a result of bacterial growth?

(a) A meat product with near neutral pH (b) Fruit in a high sugar syrup
(c) A vegetable in a high salt and brine (d) All of the above

307. Iodine sanitizers causes death of pathogens, due to-------------

(a) Destruction of mitochondria (b) Damage to DNA
(c) Cell wall damage (d) None of the above

308. A psychrotrophic non-sporeforming pathogen that can grow at refrigerator temperatures is -

(a) *Clostridium botulinum* (b) *Bacillus cerus*
(c) *Clostridium perfringens* (d) *Listeria monocytogenes*

309. A musty or muddy odour of the fish is attributed to-----------

(a) The growth of *Streptomyces* species (b) The growth of *Clostridium* species
(c) The growth of *Pseudomonas* species (d) None of the above

310. In chilled shrimp --------------- is chiefly responsible for spoilage.

(a) *Achromobacter* (b) *Pseudomonas*
(c) *Micrococcus or Bacillus species* (d) Molds or yeasts

311. Marinated (sour pickled) fish do not have spoilage problems unless--------------

(a) The acid content is very high (b) The acid content is low enough
(c) The acid content is moderate (d) None of the above

312. The predominant kind of bacteria causing spoilage in fish at chilling temperature is--------

(a) *Species of Pseudomonas* (b) *Micrococcus*
(c) *Bacillus* (d) None of the above

313. The bacteria most often involved in the spoilage of fish are------------
(a) Part of the natural flora of the external slime of fishes and their intestinal contents
(b) Part of the natural flora of the internal slime of fishes only
(c) Both a and b
(d) None of the above

314. Which of the following strain(s), that grow in the jellies and candid fruits, are able to grow in sugar concentration up to 67.5%?
(a) *Aspergillus* (b) *Penicillium* (c) *Citromyces* (d) All of the above

315. The immediate source of the flat sour bacteria is usually the-------------
(a) Plant equipment (b) Sugar (c) Starch (d) Soil

316. The ability of *B. coagulans* to grow in tomato juice depends upon the------------
(a) Number of spores present (b) The availability of oxygen
(c) The pH of the juice (d) All of the above

317. The most important kind of chemical spoilage of canned foods is the hydrogen swell which is favoured by-----------
(a) Increasing acidity of foods (b) Increasing temperature of storage
(c) Imperfection in the tinning and lacquering (d) All of the above

318. Which of the following is responsible for the spoilage of leaky cans?
(a) Plant equipment (b) Cooling water
(c) Highly acidic medium (d) High sugar content

319. Canned, sweetened condensed milk may become thickened by------------
(a) *Bacillus species* (b) *Clostridium species*
(c) *Micrococcus species* (d) *Saccharomyces species*

320. The canned acid foods have been found spoiled by *C. pasteurianum*. Such spoilage is more likely when the pH of the food is----------
(a) >4.5 (b) <4.5 (c) >5.3 (d) <5.3

321. Which of the following species can cause bitterness, acidity and curdling in canned milk, cream and evaporated milk?
(a) *Bacillus* (b) *Clostridium* (c) *Saccharomyces* (d) Both a and b

322. Acid foods with a pH between 4.5 and 3.7 usually are spoiled by-----------------
(a) *B. coagulans* (b) *Saccharolytic anaerobe*
(c) Both a and b (d) None of the above

323. The butyric acid type of fermentation with swelling of the container by the carbon dioxide and hydrogen gas is caused by------------
(a) *C. butyricum* (b) *C. pasteurianum* (c) *C. sporogenes* (d) Both a and b

324. Which of the following(s) is/are putrefactive decomposing proteins with the production of malodorous compounds such as H_2S, mercaptans, ammonia, indole and skatole?
(a) *C. sporogenes* (b) *C. putrifaciens* (c) Both a and b (d) *C. butyricum*

325. Canned poultry is more often spoiled by------------

(a) Putrefactive than by *Saccharolytic clostridia*

(b) Saccharolytic clostridia than by putrefactive

(c) Saccharomyces species

(d) Micrococcus species

326. High acid foods with a pH < 3.7 undergo spoilage by-------------

(a) *Saccharolytic anaerobe*

(b) *B. coagulans*

(c) Do not undergo spoilage by microorganisms

(d) Both a and b

327. The putrefactive anaerobes grow best in the-------------

(a) Low acid canned food (b) Medium acid canned foods

(c) High acid canned foods (d) None of the above

328. Which of the following bacteria forms endospores?

(a) *Bacillus* (b) *Clostridium* (c) *Sporolactobacillus* (d) All of the above

329. During germination, the spores releases-----------

(a) Succinic acid (b) Ascorbic acid (c) Tartaric acid (d) Dipicolinic acid

330. The organism responsible for spoilage of pasteurized milk compared to raw milk is ----------

(a) *B. coagulans* (b) *B. cereus*

(c) *B. subtilis* (d) *Lactobicillus acidophilus*

331. Microcycle includes-----------

(a) Spore- cell-spore (b) Cell-spore (c) Spore-cell (d) None of the above

332. *Bacillus* organisms produce------------

(a) Acid (b) Catalase (c) Gas (d) Both a and b

333. Which of the following spores are more sensitive to any stress?

(a) Matured (b) Old (c) Young (d) All of the above

334. Organisms are more resistant to heat at -------------pH of the medium.

(a) 7.0 (b) >8.0 (c) <6.0 (d) None of the above

335. To have reasonable assurance that spores of *C. botulinum* are destroyed ----------- concept is used.

(a) 10 D (b) 11 D (c) 12 D (d) 13 D

336. In 12 D concept in thermal processing means that the number of spores will be reduced by --------------- log cycles.

(a) 12 (b) 13 (c) 10 (d) 9

337. In 12 concept original 10 ^{6} number of C. *botulinum* spores would be reduced to ------------

(a) 10^5 (b) 10^3 (c) 10^2 (d) 10^1

338. Spores of which organism are highly resistant to heat-------------

(a) *C. perfringence* (b) *C. botulinum* (c) *B. cereus* (d) *B. stearothermophilus*

339. *Listeria monocytogenes* is --------------

(a) Gram positive and rod shaped (b) Gram negative and rod shaped
(c) Non motile (d) None of the above

340. *Listeria monocytogenes* is --------------

(a) Salt tolerant (b) Can grow at refrigeration temperature
(c) Can grow better at lower O_2 level (d) All of the above

341. Zearalenone is synthesis by ---------------

(a) *Alternaria* spp (b) *Penicillium* spp (c) *Fusarium* spp (d) All of the above

342. Which one of the mycotoxins is not produced by *Fusarium* spp.?

(a) Don (b) Nivalenol (c) T-2 (d) Sterigmatocystin

343. A cluster of polar flagella is called---------------

(a) Lophotrichous (b) Amphitrichous (c) Monotrichous (d) Petritrichous

344. Peptidoglycan accounts for ----------------- of the dry weight of cell wall in many gram positive bacteria.

(a) 50% or more (b) About 10% (c) 11%+ 0.22% (d) About 20%

345. Which of the following is true about cell wall of gram positive bacteria?

(a) It consists of multiple layers
(b) It is thicker than that associated with gram-negative bacteria
(c) It contains teichoic acids
(d) All of the above

346. The cell wall of-------------

(a) Gram positive bacteria are thicker than gram negative bacteria
(b) Gram negative bacteria are thicker than gram positive bacteria
(c) Both have same thickness but composition is different
(d) None of the above

347. Peptidoglycan is also known as--------------

(a) N-acetyl muramic acid (b) Murein mucopeptide
(c) N acetylglucosamine (d) Mesodiaminopimetic acid

348. Red wine is obtained by fermentation of----------------

(a) Banana juice (b) Red roses
(c) Grapes by red pigment producing bacteria (d) Grapes with skin

349. Gray mold rot in grapes is caused due to---------------

(a) *B. cinera* (b) *R. stolonifer* (c) *A. niger* (d) None of the above

350. Which of the following has peptidoglycan as a major constituent of cell wall?

(a) Gram negative bacteria (b) Gram positive bacteria
(c) Fungi (d) None of the above

351. Gram positive cells have a--------------
    (a) Second outer membrane that helps to retain the crystal violet stain
    (b) Multiple layer of peptidoglycan that helps to retain the crystal violet stain
    (c) Thick capsule that traps the crystal violet stain
    (d) Periplasmic space that traps the crystal violet
352. Teichoic acids are typically found in----------------
    (a) Cell walls of gram positive bacteria (b) Outer membranes of gram positive bacteria
    (c) Cell walls of gram negative bacteria (d) Outer membranes of gram negative bacteria
353. In eukaryotic cells, ribosomes are-------------
    (a) 70S (b) 60S (c) 80S (d) Not specific
354. Which of the following is exposed on the outer surface of a gram negative bacterium?
    (a) O-antigen of lipopolysaccharide (LPS)
    (b) Polysaccharide portion of lipoteichoic acid (LTA)
    (c) Braun lipoprotein
    (d) Electron transport system components
355. Which of the following does not contain protein?
    (a) Pili (b) Flagellum (c) Lipoteichoic acid (d) Porin
356. An important characteristic of mycobacteria is that they are?
    (a) Gram negative (b) Rapid growing (c) Acid fast (d) Alpha hemolytic
357. All of the following are sporicidal except-------------
    (a) Glutaraldehyde (b) Ethylene oxide (c) Formaldehyde (d) Alcohol
358. Which of the following is not a disinfectant containing heavy metal?
    (a) Silver nitrate (b) Mercurochrome (c) Chlorine (d) Copper sulfate
359. Which of the following was the first widely used antiseptic and disinfectant?
    (a) Chlorine (b) Phenol (c) Iodine (d) Alcohol
360. A psychrotrophic non-sporeforming pathogen that can grow at refrigerator temperatures is--
    (a) *Clostridium botulinum* (b) *Bacillus cerus*
    (c) *Clostridium perfringens* (d) *Listeria monocytogenes*
361. Who has developed a procedure to heat wine at temperatures well below boiling to prevent spoilage of the wine?
    (a) Pasteur (b) Lister (c) Koch (d) Watkins
362. Organisms, which utilize simple inorganic compounds, are termed as---------------
    (a) Phototrophs (b) Autotrophs (c) Heterotrophs (d) Necrotrophs
363. Selman A. Waksman was awarded Nobel Prize for the discovery of---------------
    (a) Penicilin (b) Streptomyces (c) Streptomycin (d) Penicillium

364. Robert Koch has been awarded Nobel Prize in 1905 for the discovery of----------------

(a) Anthrax (b) Cholera (c) Tuberculosis (d) None of the above

365. The group of bacteria which includes obligate intracellular parasites is------------------

(a) Rickettsias (b) Mollicutes (c) Archaeobacteria (d) Actinomycetes

366. Which of the following methods is employed for disposal of sludge that is contaminated with heavy metals?

(a) Incineration (b) Drying (c) Land filling (d) Composting

367. C. *botulinum toxin* / toxins responsible for *human* botulism is/are------------

(a) Types A (b) Type B (c) Type C (d) Both a and b

368. The most common symptom of a *Campylobacter* infection is -----------------

(a) Diarrhea, which is often bloody (b) Headache

(c) Muscle pain (d) Excess salivation

369. *Campylobacter* infection is more common in which group(s) of people?

(a) People who travel to developing countries where sanitation and food hygiene may be less strict

(b) People working with farm animals

(c) People who work in the meat industry

(d) All of the above

370. Verocytotoxin is produced by---------------

(a) *E. coli* O157 (b) *Salmonella typhi* (c) *Shigella* spp (d) *Campylobacter jejuni*

371. Sulfur stinker spoilage is uncommonly found in------------------

(a) Low acid foods (b) Medium acid foods (c) High acid foods (d) All of the above

372. Medium acid foods with a pH between 5.3 and 4.5 are likely to undergo-----------------

(a) Flat sour spoilage (b) Putrefaction

(c) TA spoilage (d) Both a and b

373. Decimal reduction time (D) is the time required to reduce the microbial population by ----------------- % at specified temperature.

(a) 90 (b) 10 (c) 20 (d) 30

374. D value in death rate curve indicates time required for survival or death rate curve to traverses------------------

(a) One log cycle (b) 1/10 log cycle (c) 1/100 log cycle (d) None of the above

375. Some breweries, formation of hot break is enhanced by addition of----------------

(a) Calcium sulphate (b) Lactic acid (c) Tannic acid (d) None of the above

376. The disease which spreads from one continent to other is termed as-------------

(a) Pandemic (b) Sporadic (c) Epidemic (d) None of the above

377. Sour flavour in milk during storage is caused by--------------

(a) *Alkaligenes viscolactis* (b) *Streptococcus lactis*

(c) *Pseudomonas suncyanea* (d) *Bacillus subtilis*

378. The cell walls of gram positive bacteria contain two modified sugar, viz. N-acetylglucosamine (NAG) and N- acetylmuramic acid (NAM). They are covalently linked by--------------

(a) α- 1,4-glycosidic bond (b) β-1,6-glycosidic bond

(c) α- 1,6-glycosidic bond (d) β- 1,4-glycosidic bond

379. In preparation of curd from milk the fermentation is terminated due to-------------

(a) Complete oxidation of milk sugar (b) Death of fermenting organisms

(c) Fall in pH which inhibit microbial activity (d) None of the above

380. Beneficial bacteria generally come from which lactic acid species?

(a) *Bifidobacterium* (b) *Enterococcus* (c) *Lactobacillus* (d) All of the above

381. A culture started with 4 cells and ended with 128 cells. How many generations did the cells go through?

(a) 64 (b) 32 (c) 6 (d) 5

382. Typical temperature for an autoclave (operating at 15 pounds per square inch of pressure) is ------------°C.

(a) 121 (b) 100 (c) 63 (d) 73

383. In recent years, the potential for food contamination has been increased by-------------

(a) Multiple handling of food in supply chain from production to final customer

(b) Desire of customers to experiment with thoroughly cooked foods

(c) Decrease of overall quality of foods in the United States

(d) Lack of federal, state and local government regulation of food products

384. The micro-organisms of greatest concern to foodservice managers are--------------

(a) Parasites (b) Molds (c) Fungi (d) Bacteria

385. Which of the the following is an example of a food borne disease-causing parasite?

(a) *Canpylobacter jcjuni* (b) *Salmonella enteritidus*

(c) *Clostridium perfringens* (d) *Trichinella spiralis*

386. In maintaining a course of good personal hygiene rnanagement most important responsibility is to--------------

(a) Provide hand soap at all sinks

(b) Set an example for all employees to follow

(c) Yell at any employees who do not follow the rules

(d) Provide a designated break area for employees

387. Which of the following describes the proper way to dry hands after they have been washed?

(a) Thoroughly with a hot air hand dryer or a single service towel

(b) Wipe them vigorously on an apron or a handkerchief

(c) Use a cloth towel that is kept in the restroom

(d) Wave hands briskly back and forth

388. For amylase production---------------- microorganism is required.
(a) *B. subtilis* (b) *S. cereviceae* (c) *A. nigar* (d) None of the above

389. Pectinase is industrially produced from-----------
(a) *S.cereviceae* (b) *Trichoderma koningi*
(c) *A. nigar* (d) None of the above

390. Cellulose are produced from--------------
(a) *S.cereviceae* (b) *Trichoderma koningi*
(c) *A. nigar* (d) None of the above

391. Industrial production of vitamin B_{12} is from--------------
(a) *Propionibacterium sps.* (b) *Pseudomonas sps.*
(c) Both a and b (d) None of the above

392. *Clostridium acetobutylicum* is used for the production of--------------
(a) Acetone – Butanol (b) Ethanol
(c) Vitamin B_{12} (d) None of the above

393. Which yeast is used in the industrial production of ethanol?
(a) *K.pneumoniae* (b) *Kluyreromyces fragilis*
(c) *S. cerevisiae* (d) Both b and c

394. Citric acid is produced in aerobic conditions by which fungi?
(a) *Aspergillus* (b) *Penicillin* (c) *Mucor* (d) All of the above

395. Cleaning anodized aluminum containers require the use of detergents that are having--------
(a) Acid pH (b) Neutral pH (c) Alkaline pH (d) Highly alkaline pH

396. What are the signs of a food that has been contaminated by *Clostridium botulinum*?
(a) Bulging, leaking, or severely dented cans that have cracks and bulging lids
(b) A gray-green coloured food upon exposure to air
(c) A sour taste or smell
(d) A foul aroma

397. The foods most often contaminated with aflatoxins are--------------
(a) Cheeses made from unpasteurized milk (b) Meats from wild animals such as bear and deer
(c) Fish and shellfish (d) Peanuts and tree nuts

398. Cross-contamination of foods occurs --------------
(a) When perishable foods are kept at room temperature up to two hours
(b) When two or more food handlers are working with the same food
(c) When two or more different microorganisms are growing in the same food
(d) When a utensil contaminated with a microorganism from a previously handled food is allowed to come in contact with a second food and contaminates it

399. In alchoholic fermentation, CO_2 is evolved during---------------

(a) Decarboxylation of pyruvic acid (b) Formation of acetaldehyde

(c) Oxidation of acetaldehyde (d) Both a and b

400. In the industrial production of streptomycin, the secondary metabolite is--------------

(a) Vitamin B_{12} (b) Vitamin C (c) Vitamin B_6 (d) Ethanol

401. Tobacco and tea leaves are fermented to give flavour and taste. This type of fermentation is known as---------------

(a) Alcohol fermentation (b) Curing

(c) Degradation (d) Lactic acid fermentation

402. Vinegar fermentation involves------------

(a) Yeasts only (b) Yeasts with lactic bacteria

(c) Yeasts with acetic acid bacteria (c) Yeasts with butric acid bacteria

403. Thermo resistant bacteria are important in the preservation of foods by-------------

(a) Freezing (b) Canning (c) Chemicals (d) Irradiation

404. The fungus used in the industrial production of citric acid is------------

(a) *Rhizopus Oryzae* (b) *Fusarium moniliformae*

(c) *Rhizopus nigricans* (d) *Aspergillus nigricans*

405. Penicilin is commercially produced by---------------

(a) *P.notatum* (b) *P.chrysogenum* (c) *P.citrinum* (d) *P.roquefortii*

406. --------------- is a toxic element which may contaminate food.

(a) Zinc (b) Copper (c) Lead (d) Calcium

407. The toxin found in groundnut kernel is---------------

(a) Aflatoxin (b) Patulin (c) Aminotoxin (d) Paratoxin

408. Decomposition of carbohydrates by microorganisms or enzymes is called as-----------

(a) Putrefaction (b) Fermentation (c) Canning (d) Dextrinization

409. Bacteria that cause food spoilage---------------

(a) Are a type of enzyme (b) Are only present in chicken

(c) Will give the food an 'off' odour (d) Need a moist, damp environment to grow

410. Which medium is used for the production of penicillin using immobilized cells?

(a) 1% peptone (b) Glucose (c) Yeast extract (d) LB broth

411. Which microbial culture is widely used for fermentation of fruit juices into wine?

(a) *Schizosaccharomyces*

(b) *Saccharomyces uvarum*

(c) *Saccharomyces cerevisiae var. ellipsoideus*

(d) *Z. mobilis*

412. In Tristimulus colorimetry Z stands for---------------
(a) Green (b) Blue (c) Amber (d) Red

413. The UV spectrum has wavelengths between--------------
(a) 50-100 nm (b) 100-400 nm (c) 400-700 nm (d) None of the above

414. Which acid is produced as a result of activity of insects and rodents during the storage of farm produce?
(a) Oxalic acid (b) Uric acid (c) Citric acid (d) Malic acid

415. Vitamins, Minerals and Proteins in suitable amounts are given to body by-----------
(a) Fatty fish (b) Whole grain food (c) Nuts (d) Plant oil

416. The disease hyponatremia is caused due to ------------
(a) Higher potassium level (b) Higher sodium level
(c) Lower potassium level (d) Lower sodium level

417. Which of the following is not energy yielding nutrient?
(a) Vitamin (b) Protein (c) Carbohydrate (d) Fat

418. Which is not essential in a balanced diet normally?
(a) Fruits and vegetables (b) Salads
(c) Pulses (d) Tea

419. Food that provides protection against diseases is/are------------
(a) Sugar (b) Ghee (c) Potatoes (d) Green leafy vegetables

420. Haemoglobin formation needs both------------
(a) Iron and zinc (b) Iron and calcium (c) Iron and folic acid (d) Iron and ascorbic acid

ANSWER KEY

1. a	2. a	3. a	4. a	5. a	6. a	7. c	8. b	9. c	10. c
11. b	12. c	13. a	14. b	15. d	16. a	17. b	18. a	19. b	20. a
21. a	22. b	23. a	24. c	25. a	26. d	27. a	28. d	29. a	30. d
31. d	32. b	33. b	34. b	35. b	36. a	37. b	38. a	39. c	40. a
41. c	42. a	43. a	44. c	45. b	46. a	47. b	48. a	49. a	50. b
51. a	52. a	53. b	54. c	55. b	56. c	57. b	58. d	59. b	60. a
61. c	62. a	63. a	64. b	65. c	66. a	67. c	68. b	69. a	70. c
71. b	72. a	73. d	74. a	75. b	76. a	77. a	78. c	79. a	80. c
81. d	82. a	83. c	84. d	85. b	86. a	87. c	88. b	89. c	90. c
91. b	92. d	93. a	94. a	95. c	96. b	97. d	98. c	99. a	100. c
101. b	102. a	103. b	104. b	105. a	106. c	107. a	108. c	109. d	110. b
111. d	112. a	113. c	114. b	115. d	116. a	117. a	118. c	119. a	120. a
121. a	122. d	123. d	124. a	125. a	126. c	127. b	128. a	129. b	130. a

131. a	132. a	133. a	134. b	135. b	136. d	137. c	138. d	139. d	140. a
141. d	142. c	143. b	144. d	145. a	146. a	147. c	148. b	149. b	150. d
151. c	152. b	153. a	154. c	155. a	156. a	157. b	158. a	159. c	160. d
161. c	162. c	163. d	164. c	165. b	166. c	167. a	168. a	169. b	170. c
171. b	172. c	173. a	174. c	175. d	176. b	177. d	178. a	179. a	180. d
181. a	182. a	183. a	184. d	185. c	186. a	187. c	188. b	189. c	190. d
191. b	192. a	193. c	194. a	195. c	196. b	197. d	198. a	199. a	200. b
201. a	202. a	203. b	204. b	205. d	206. a	207. a	208. d	209. b	210. c
211. c	212. d	213. d	214. a	215. d	216. c	217. d	218. d	219. a	220. d
221. b	222. a	223. a	224. a	225. d	226. a	227. c	228. c	229. c	230. c
231. a	232. a	233. a	234. a	235. d	236. c	237. c	238. a	239. a	240. c
241. a	242. c	243. a	244. a	245. d	246. b	247. c	248. a	249. a	250. b
251. b	252. b	253. b	254. b	255. c	256. a	257. a	258. c	259. d	260. a
261. a	262. c	263. b	264. d	265. d	266. a	267. c	268. d	269. b	270. d
271. a	272. b	273. d	274. c	275. a	276. d	277. a	278. c	279. b	280. b
281. d	282. a	283. c	284. c	285. b	286. b	287. c	288. b	289. a	290. b
291. a	292. b	293. a	294. b	295. c	296. c	297. a	298. d	299. d	300. c
301. b	302. d	303. b	304. c	305. b	306. a	307. c	308. d	309. a	310. a
311. b	312. a	313. a	314. d	315. a	316. d	317. d	318. a	319. c	320. a
321. b	322. c	323. d	324. c	325. a	326. c	327. a	328. d	329. d	330. b
331. a	332. d	333. c	334. a	335. c	336. a	337. a	338. d	339. a	340. d
341. c	342. d	343. a	344. a	345. d	346. a	347. b	348. d	349. a	350. b
351. b	352. a	353. c	354. a	355. c	356. c	357. d	358. c	359. b	360. d
361. a	362. b	363. c	364. a	365. a	366. d	367. d	368. a	369. d	370. a
371. a	372. c	373. a	374. a	375. c	376. a	377. b	378. d	379. c	380. d
381. d	382. a	383. a	384. d	385. d	386. b	387. a	388. a	389. c	390. b
391. c	392. b	393. d	394. d	395. b	396. a	397. d	398. d	399. d	400. a
401. b	402. c	403. b	404. d	405. b	406. c	407. a	408. b	409. d	410. b
411. c	412. b	413. b	414. b	415. c	416. d	417. a	418. d	419. d	420. c

5

Food Business Management

1. Working together for promotion of economic interest means ------------ movement.
   (a) Capitalism (b) Communism (c) Co-operation (d) None of the above
2. ICA stands for-------------
   (a) International Communism Alliance (b) International Capitalism Association
   (c) International Co-operative Alliance (d) None of the above
3. Commission on Agricultural Costs and Prices (CACP) was established in which year?
   (a) 1960 (b) 1965 (c) 1970 (d) 1975
4. CACP has been operating which set of prices?
   (a) Procurement price (b) Minimum support price
   (c) Both a and b (d) None of the above
5. NAFED was established in-------------
   (a) 1958 (b) 1956 (c) 1960 (d) 1962
6. FCI was established in-------------
   (a) 1961 (b) 1963 (c) 1965 (d) 1967
7. Govt. announces which of the following price well in advance of sowing season of the crop?
   (a) Minimum support price (b) Procurement price
   (c) Both a and b (d) None of the above
8. Which of the following word denote the capacity of borrower to borrow?
   (a) Credit (b) Finance (c) Loan (d) None of the above
9. In financial analysis statement, ratio of total current assets to total current liabilities is known as ------------
   (a) Current profit (b) Current ratio (c) Current finance (d) Current credit
10. Which committee recommends establishing NABARD?
    (a) Shivaraman (b) Chakravarti (c) Chellaya (d) Venkatraghavan
11. Which agency is working as main authority to regulate all the banks and finance in India?
    (a) Bank of India (b) SBI (c) RBI (d) Central Bank of India

12. Food Safety and Standards Acts is promulgated in which year?
(a) 2004 (b) 2006 (c) 2008 (d) 2010
13. SWMA stands for------------
(a) Standards of Weights and Measure Act, 1976
(b) Standards of Water and Monitoring Act, 1976
(c) Standards of Water and Management Act, 1976
(d) Standards of Weights and Monitoring Act, 1976
14. Which act is related with Grading and Marketing of Agricultural Commodities?
(a) AMGARK (b) AGMARK (c) AGIMARK (d) MARCAG
15. BIS stands for-----------
(a) Bureau of Indian Standards (b) Bureau of International Spices
(c) Bureau of International Standards (d) None of the above
16. Which order is related to meat products?
(a) MMPO (b) MFPO (c) FMPO (d) None of the above
17. Provision of Food safety is made compulsory by which act?
(a) PFA (b) FPO (c) BIS (d) FSSA 2006
18. Food inspector is related with which act?
(a) PFA (b) FSSA (c) FPO (d) None of the above
19. For false advertisement, punishment / penalties are mentioned in which act?
(a) FSSA 2006 (b) PFA 1954
(c) BIS 1986 (d) Consumer Protection Act 1986
20. ---------- is the set of marketing tools that the firm uses to pursue its marketing objectives.
(a) Marketing strategy (b) Marketing mix
(c) Marketing activities (d) Marketing Process
21. A situation where countries export a product at a price below the cost of its production is known as-------------
(a) Price discrimination (b) Dumping
(c) Protectionism (d) Price skimming
22. The three pillars of WTO negotiations in agriculture are--------------
(a) Market access, domestic support and export competition
(b) Market access, antidumping and export restraints
(c) Market access, domestic subsidies and export subsidies
(d) Market access, blue box support and export credits
23. ------------- developed fourteen general principles of management.
(a) Frederick Taylor (b) Koontz (c) Henry Fayol (d) Elton Mayo

24. The Reserve Bank of India was established in-------------
(a) 1935 (b) 1953 (c) 1988 (d) 1920
25. The difference between the value of commodity exports and imports is known as ------------
(a) Balance of payments (b) Balance of trade
(c) Trade deficit (d) Terms of trade
26. An entrepreneur starts new enterprises taking risk for the purpose of achieving profit and growth by identifying---------------
(a) Products (b) People (c) Opportunities (d) Jobs
27. Products are sold directly to the people who use or consume them in-----------
(a) Wholesaling (b) Retailing (c) Service (d) Manufacturing
28. A random sample is one where ------------
(a) The researcher works to a specific number of individuals
(b) Every member of the population is selected for consultation
(c) The researcher selects an appropriate number of people to contact
(d) Every member of the population must have an equal chance of selection
29. The primary motivating force behind entrepreneurs is------------
(a) Money (b) Fame (c) Achievement (d) Recognition
30. The purpose of SWOT analysis is to analyse--------------
(a) The business environment in which an organization operates
(b) The strategic capability of an organization
(c) The business environment and the strategic capability of an organization relative to its competitors
(d) External and organizational environments
31. ------------- is a mass and paid communication of goods, services or ideas by an identified sponsor.
(a) Personal selling (b) Sales promotion (c) Public relations (d) Advertising
32. What is the motto of cooperation?
(a) Each for all and all for each (b) Each for one and all for each
(c) Each for all and all for one (d) Each for each and all for all
33. When the idea of cooperative movement was first conceived in India?
(a) 1890 (b) 1892 (c) 1889 (d) 1888
34. A market which is located in a small village, where major transactions take place among the buyers and sellers of a village is called as------------market.
(a) Village (b) Super (b) Forward (d) Periodic
35. The quantity of a product or services which the buyers are likely to purchase at different prices in a given market at a given time is called as--------------
(a) Supply (b) Demand (c) Cost (d) Utility

36. Any name, sign, symbol and /or design used to identify the product of one firm which set them apart from competitor's offerings is known as-------------
(a) Brand (b) Image (c) Identity (d) Model

37. The difference between the value of exports and imports is called as-------------
(a) Excise duty (b) Balance of trade (c) Export import value (d) International trade

38. An association of firms that attempts to regulate industry output and prices through mutual agreement is termed as-------------
(a) Franchise (b) Cartel (c) Company (d) Organization

39. The marketing technique of developing separate products and marketing programmes to appear to different consumer classes is called as-------------
(a) Market Mix (b) Market segmentation
(c) Marketing strategy (d) Market positioning

40. The sequence of events and actions that coordinate the flow of goods and the value adding activities in the marketing system means-------------
(a) Marketing process (b) Market mix
(c) Market channel (d) Market flow

41. The volume of sale at which there is no profit or loss is known as-------------
(a) Sales target (b) Breakeven point (c) Profit point (d) Sale

42. A process by which a cooperative group directs action towards common goals is known as---
(a) Management (b) Organization (c) Corporate (d) Firms

43. What is GATT?
(a) General Agreement on Tariffs and Trade (b) Goods Agreement on Trade and Tariffs
(c) General Agreement on Taxes and Trade (d) General Agriculture of Trade and Tariffs

44. Head Quarter of tea board is located at------------
(a) Mumbai (b) Kolkata (c) Chennai (d) Delhi

45. ---------------is a legal term given to particular branded product duly registered under the trade names and trademarks act.
(a) Trade Mark (b) Terms of Trade (c) Brand Name (d) Quality Mark

46. Quantity of the produce which the producer actually sells in the market is known as----------
(a) Marketable surplus (b) Marketed surplus
(c) Market value (d) Selling surplus

47. Head Quarter of Spices board is located at-------------
(a) Cochin (b) Kolkata (c) Mumbai (d) Delhi

48. Head Quarter of Coffee board is located at-------------
(a) Jaipur (b) Anand (c) Mysore (d) New Delhi

49. Canadian and American laws prescribe how ingredients should be listed on labels. What is the general stipulation with respect to the order that ingredients are listed?

(a) By alphabetical order
(b) By ascending order of proportion by weight
(c) By descending order of proportion by weight
(d) By descending order of proportion by volume

50. Which of the following are functions of management?
(a) Planning and directing (b) Discipline and hierarchy
(c) Orderliness and work stability (d) Confucian and ethics

51. Which of the following are the barriers to communication?
(b) Bad memory and poor listening (b) Leading and staffing
(c) Good language and planned message (d) Guiding and appreciation

52. The exchange of goods between citizens of different countries is called as-------------
(a) Internal trade (b) International trade (c) Domestic trade (d) Commodities trade

53. Rate at which a unit of one country exchanges for the currency of another is known as---------
(a) Rate of exchange (b) Foreign payment (c) Exchange control (d) Balance of payment

54. IMF means----------
(a) Indian Marketing Federation (b) International Monetary Fund
(c) Indian Monetary Fund (d) International Marketing Federation

55. Policy of no restrictions in the movement of goods between countries is known as------------
(a) Free market (b) Free trade (c) Protection (d) Bounties

56. The minimum support price are declared by the government after-------------
(a) Every two years (b) Every year (c) Every six month (d) Every five years

57. The three major functions of business organizations are--------------
(a) Are mutually exclusive (b) Exist independently of each other
(c) Overlap (d) Function independently of each other

58. Which of the following is an area approach for lead bank?
(a) Village (b) District (c) Tehsil (d) State

59. Which is the leading lending agency at international level?
(a) RBI (b) World Bank (c) NABARD (d) AFC

60. ------------ protects the farmers from heavy losses of crops yields.
(a) Crop insurance scheme (b) Minimum support price
(c) NAFED (d) NABARD

61. Which committee recommended setting up Regional Rural Banks (RRBS)?
(a) Narasimhan working groups (b) Vyas committee
(c) Maclagan committee (d) None of the above

62. ------------ performs major role in monetary stability in the country.
(a) RBI (b) AFC (c) NABARD (d) ADB

63. International Organization for Standardization (ISO) comes in to existence in ------------
(a) 1958 (b) 1947 (c) 1950 (d) 1961
64. How much refinance is provided by NABARD for agro processing units?
(a) 25% (b) 50% (c) 60% (d) 75%
65. Which of the following belong to the tertiary function of market?
(a) Processing (b) Transportation (c) Buying and Selling (d) Communication
66. Which of the following does not concern with creation of storage facilities?
(a) CWC (b) SWC (c) FCI (d) APEDA
67. The head office of NAFED is located at------------
(a) Mumbai (b) Chennai (c) Nagpur (d) New Delhi
68. APEDA was established on---------------
(a) February 13, 1986 (b) August 9, 1955
(c) January 26, 1947 (d) February 13, 1983
69. MPEDA was set up to undertake the promotional work relating to the export of-----------
(a) Milk products (b) Meat products (c) Marine products (d) All of the above
70. Which is/ are the commodity board?
(a) Tea board (b) Coffee board (c) Spices board (d) All of the above
71. CCFS stands for-------------
(a) Central Council of Food Safety (b) Central Committee for Food Standards
(c) Central Committee for Food Storage (d) None of the above
72. Which is the co-operative marketing federation?
(a) RAJFED (b) MARKFED (c) NAFED (d) All of the above
73. World Trade Organization is an international body to supervise and encourage-----------
(a) National Trade (b) International Trade (c) Both a and b (d) None of the above
74. What is WTO?
(a) World Trade Organization (b) World Textile Organization
(c) World Testing Organization (d) World Transportation Organization
75. Full form of SEZ is ------------
(a) State Export Zone (b) Special Economic Zone
(c) Safety Economic Zone (d) Sanctioned Economic Zone
76. To ensure availability of right people for right job is the primary objective of-----------
(a) Production Management (b) Marketing Management
(c) Human Resource Management (d) Financial Management
77. Head office of APEDA is located at ------------
(a) Mumbai (b) Kolkata (c) Hyderabad (d) New Delhi

78. Trade related intellectual property rights include------------

(a) Copyright and trade-marks (b) Geographic indication

(c) Industrial design and patents (d) All of the above

79. EXIM means--------------

(a) Exclude and Include Markets (d) Extra Income Management

(c) Export–Import (d) Export Incoming Market

80. International trade essentially involves buyers and sellers between-------------

(a) Two different states (b) Two different countries

(c) Only a (d) Both a and b

81. Theory of absolute advantage was first explained by-----------

(a) Adam Smith (b) John Henry (c) David Ricardo (d) David Smith

82. David Ricardo, a classical economist developed the principle of ------------

(a) Equal Cost Advantage (b) Comparative Cost Advantage

(c) Absolute Cost Advantage (d) All of the above

83. Which theory of international trade assures barter system of exchange?

(a) Classical Theory (b) Modern Theory (c) Executive Theory (d) Both a and b

84. Full form of MOFPI is ----------------

(a) Maharashtra's Organization for Food and Integrated Products

(b) Maharashtra Food Processing Industries

(c) Maharashtra Food and Pulses Industries

(d) Ministry of Food Processing Industries

85. Commodity boards are controlled organization by--------------

(a) Consumers (b) Producers (c) Ministry (d) None of the above

86. Productivity of labour is given by--------------

(a) Productivity of labour = Output / man hrs. spent

(b) Productivity of labour = Output / machine hrs. works

(c) Productivity of labour> Output / man hrs. spent

(d) None of the above

87. A formal agreement between states for the purpose of establishing mutual rights and regulating conditions of trade is----------------

(a) Production Treaty (b) Commercial Treaty

(c) Financial Treaty (d) Operational Treaty

88. The act of ---------------was a pioneer act in the cooperative credit movement in India.

(a) 1912 (b) 1901 (c) 1904 (d) 1905

89. The Prevention of Food Adulteration Act was established in-------------

(a) 1953 (b) 1954 (c) 1955 (d) 1956

90. The term Marketing Mix was coined in 1953 by--------------
(a) David Recardio (b) Neil Borden (c) Adams Smith (d) None of the above

91. The element of the marketing mix is often referring to as--------------
(a) Four M (b) Four P (c) Four C (d) Both b and c

92. WTO began functioning on----------------
(a) 15th April 1994 (b) 1st Jan. 1948 (c) 1st Jan 1995 (d) 1st April 1999

93. Press, advertisement are the source of--------------
(a) Recruitment within organization (b) Recruitment outside organization
(c) Both a and b (d) None of the above

94. The absolute advantage which determine by simple comparison of roductivity of --------------
(a) Material (b) Labour (c) Land (d) None of the above

95. The regulation of International Trade is done through--------------
(a) WTO (b) MERCOSUR (c) GATT (d) NAFTA

96. ---------------- are specific limit on the quantity or value of goods that can be imported or exported during the specific time period.
(a) Tarriff quota (b) Quantitative restriction
(c) Qualitative restriction (d) Anti-dumping

97. Higher productivity is related to--------------
(a) Efficiency of production (b) Efficiency of utilization of resources
(c) Efficiency of human resource (d) None of the above

98. Any organization is depending upon-------------formula
(a) 6M (b) 5M (c) 4M (d) 3M

99. Which is the powerful tool or instrument in production management?
(a) Staffing (b) Planning (c) Controlling (d) Coordination

100. FPO stands for------------
(a) Fruit Products Order (b) Fruit Production Organization
(c) Fruit Prevention Order (d) None of the above

101. -------------- is the export promotion institution.
(a) APEDA (b) CFTRI (c) DFRL (d) CCPI

102. The main standard setting body in India is--------------
(a) BIS (b) PFA (c) MFPI (d) None of the above

103. HACCP is---------------
(a) Scientific way of enhancing the safety of food
(b) A tool to access hazards
(c) Enhances the safety of end product only
(d) Both a and b

104. Codex Alimentarius Commission is run by--------------
(a) FAO (b) WHO (c) FAO and WHO (d) ISO

105. HACCP technique is applied for control measures associated with--------------
(a) Quality (b) Safety (c) Safety and quality (d) None of the above

106. HACCP is --------------
(a) Recommendation given by CAC (b) Specifications given by CAC
(c) Standards given by CAC (d) All of the above

107. A market having more than two buyers is known as--------------market.
(a) Oligopsony (b) Duopsony (c) Oligopoly (d) Duopoly

108. ------------ is a market situation in which there is only one seller of the commodity.
(a) Duopoly (b) Monopoly (c) Monopsony (d) Duopsony

109. Storage function adds ------------------ utility to the products by making them available at the time when they are needed.
(a) Place (b) Time (c) Form (d) Possession

110. In ------------ method, the produce is sold under cover, as one lot.
(a) Dara sale (b) Moghum sale (c) Open auction (d) Hatha sale

111. ------------ includes information relating to the facts of the prices that prevailed in the past and market arrivals over time.
(a) Market news (b) Market information (c) Market intelligence (d) Market data

112. Speculation and hedging are the important ways of minimizing the ------------ risks in business.
(a) Price (b) Product (c) Institutional (d) Personal

113. ------------ refers to the purchase or sale of commodity in a futures market accompanied by a sale or a purchase in the cash market.
(a) Hedging (b) Speculation proper
(c) Illegitimate speculation (d) Terms of trade

114. ---------------- maintains buffer stock of food grains for nation.
(a) State Govt. (b) Private sector (c) APM C (d) FCI

115. Agricultural produce (Grading and Marking Act 1937) act defines--------------
(a) Purchase (b) Sales (c) Storage (d) Quality by grade

116. ------------ includes store keeping, records keeping, maintenance of materials in store and issue of materials.
(a) Purchasing section (b) Store management
(c) Quantity control section (d) Receiving section

117. Production cycle starts and ends with the--------------
(a) Producer (b) Customer (c) Retailer (d) Wholesaler

118. Difference between consumer's price and producer's price is known as--------------
(a) Price spread (b) Marketing cost (c) Marketing margin (d) All of the above

119. Method of open auction sale of agricultural produce is followed in --------------market.
(a) Regulated (b) Weekly (c) Monopoly (d) None of the above

120. The price declared by the Government to safeguard the interest of farmers during bumper production is called as -----------------price.
(a) Minimum support (b) Procurement
(c) Market (d) All of the above

121. Maharashtra state Agricultural Marketing Board is situated at-----------
(a) Mumbai (b) Kolhapur (c) Nagpur (d) Pune

122. The Primary co-operative credit societies is a part of--------------structure.
(a) Two-tier (b) Four-tier (c) Three-tier (d) None of the above

123. Sales promotion is ------------term incentive to increase sale.
(a) Long (b) Short (c) Mid (d) None of the above

124. All the activities involved in selling goods and services to those buying for resale or business are governed by --------------
(a) Retailer (b) Wholesaler (c) Consumer (d) Merchant

125. The process that guides the implementations of activities specified in strategic plan is called as-------------- planning.
(a) Strategic (b) Tactical (c) Economical (d) Layout

126. The price at which demand and supply are equal known as an---------------price.
(a) Equilibrium (b) Market (c) Demand (d) General

127. Regulated market is the classification of market on the basis of---------------
(a) Stage of marketing (b) Volume of transaction
(c) Public intervention (d) Size of market

128. The first cold storage was established in India as early as in 1892 at--------------
(a) Mumbai (b) Madras (c) Chennai (d) Kolkatta

129. Regulated markets were established by---------------
(a) Central Government (b) Private sector
(c) APEDA (d) State Government

130. Consumer protection act was passed in --------------
(a) 1980 (b) 1986 (c) 1990 (d) 1985

131. The apex organization of marketing co-operatives is --------------
(a) FCI (b) SWC (c) AGMARK (d) NAFED

132. The price of eggs is decided by-------------
(a) NECC (b) NABARD (c) MAFCO (d) STC

133. ISI has been renamed as the BIS with effect from 1st April,------------
(a) 1987 (b) 1985 (c) 1988 (d) 1990

134. Modernization of industries helps in improving--------------
(a) Price (b) Marketing (c) Finance (d) Quality of produce

135. Central Food Technological Research Institute is located at---------------
(a) Hyderabad (b) Nagpur (c) Bangalore (d) Mysore

136. The coefficient of variability between duplicates is known as---------------
(a) Accuracy (b) Viscosity (c) Consistency (d) Precision

137. The need strategy is based on-------------
(a) Product and sale
(b) Consumer and market
(c) Market and sale
(d) Consumer and product

138. ---------- is first stage in product development process.
(a) Corporate decision
(b) Business analysis
(c) Product design
(d) Market testing

139. ------------- is a process designed to test the viability of products in the market.
(a) Idea generation
(b) New product development
(c) Costing and economics
(d) Product launching

140. A market survey is important because it tells you --------------
(a) How many and what types of people are shopping at a store
(b) What the customer will buy
(c) How much the customer will pay for the product
(d) All of the above

141. What is an entrepreneur?
(a) Someone who invests time and money to start a business
(b) Someone who makes a lot of money
(c) Someone who takes a risk to make a profit
(d) Both a and c

142. A message can only be deemed effective when it is ---------------
(a) Delivered with confidence
(b) Communicated face to face
(c) Understood by others and produces the intended results
(d) Repeated back as proof of understanding

143. ------------- is a scheme of Promotion and Handholding of micro and small enterprises.
(a) PMEGP (b) MSE-CDP (c) RGUMY (d) MUY

144. Entrepreneurs make the economy more ----------------
(a) Convenient (b) Compact (c) Concrete (d) Competitive

145. Which is alternatively called corporate venturing?
(a) Entrepreneurship
(b) Offering new products by an existing company
(c) Intrapreneurship
(d) Act of starting a new venture

146. The term entrepreneur came from-----------------

(a) Latin (b) French (c) UK (d) None of the above

147. Process or cycle of introducing a new product into the market is called as-------------

(a) Innovation (b) Commercialization (c) Invention (d) Imitation

148. Which is not an important function of entrepreneurship from a business studies perspective?

(a) Innovation and change (b) New venture creation

(c) Enrichment of life (d) Business growth

149. When superiors provide directions to their subordinates regarding what to do, it is known as ----------------- communication.

(a) Horizontal (b) Downward (c) Upward (d) Lateral

150. The theory of the maximum profit location is proposed by-----------------

(a) Fetter (b) Hotelling (c) August Losch (d) None of the above

151. Which of the following is usually the first step in the process of development of a new product ?

(a) Generating idea (b) Developing a prototype

(c) Listing product specification (d) Conducting feasibility study

152. FSSAI consist of--------------- members.

(a) 22 (b) 11 (c) 33 (d) 44

153. FSSAI regulates and monitors the safety of food to ensure-----------

(a) Manufacture and processing of food products

(b) Distribution, sale and import of food products

(c) Prices and taxes of food products

(d) Both a and b

154. As per FSSA, 2006, after receiving the sample the food analyst should submit the analysis report within ---------------days.

(a) 10 (b) 12 (c) 14 (d) 16

155. In FSSAI, out of 22 members------------- shall be women members.

(a) 1/5 (b) 1/4 (c) 1/3 (d) 1/2

156. Which chapter of FSSA 2006 deals with general principles of food safety?

(a) I (b) II (c) III (d) IV

157. The functions of Designated Officer at district level is----------

(a) To issue or cancel the license of food business operator

(b) To prohibit the sale of article which is in contravention of the provision in act

(c) To receive and get samples analyzed

(d) All of the above

158. Maximum penalty under FSSA 2006 for misbranded samples is---------

(a) Less than 4 lakh (b) 4 lakh

(c) Can be extended of 3 lakh (d) None of the above

159. When consumption of food results in death of the customer, the provision of penalty under FSSA 2006 is---------------

(a) Imprisonment not less than 7 years (b) Can be extended for life imprisonment
(c) Fine not less than 10 lakh (d) All of the above

160. ISO 22000 standards are associated with------------

(a) Safety and quality (b) Quality
(c) Safety (d) None of the above

161. The five elements in the management process are ---------------

(a) Plan, direct, update, lead, and supervise
(b) Accounting/finance, marketing, operations, and management
(c) Plan, lead, organize, manage, and control
(d) Plan, organize, staff, lead, and control

162. Which of the following are the primary functions of all organizations?

(a) Operations, marketing and human resources
(b) Marketing, human resources and finance/accounting
(c) Sales, quality control and operations
(d) Marketing, operations and finance/accounting

163. Which of the following is a mineral fibre?

(a) Glass (b) Asbestos (c) Nylon (d) Acetate

164. Which of the following factors does not influence BMR in a person?

(a) Age (b) Physical activity
(c) Activity of thyroid gland (d) Height

165. A diet containing right amount of energy, carbohydrates, proteins, fats, fibre, vitamins, minerals and water to flufill requirement of body is called----------

(a) Nutrition (b) Balanced diet (c) Perfect diet (d) Food pyramid

166. Cyanocobalamin is the name of which vitamin?

(a) Vitamin B_3 (b) Vitamin B_5 (c) Vitamin B_{12} (d) Vitamin B_7

167. --------------is essential for utilization and retention of calcium and phosphorous.

(a) Vitamin A (b) Vitamin C (d) Vitamin D (d) Vitamin K

168. --------------enzyme is required for digestion of fat.

(a) Rennin (b) Amylase (c) Lipase (d) Pepsin

169. Pakoras are fried, then idlies are-----------

(a) Steamed (b) Roasted (c) Baked (d) Toasted

170. Bitots spot is deficiency of-------------

(a) Iodine (b) Vitamin B6 (c) Vitamin 'A' (d) Calcium

ANSWER KEY

1. c	2. c	3. b	4. c	5. a	6. c	7. a	8. a	9. b	10. a
11. c	12. b	13. a	14. b	15. a	16. b	17. d	18. a	19. a	20. b
21. b	22. c	23. c	24. a	25. a	26. c	27. b	28. d	29. c	30. c
31. d	32. a	33. b	34. a	35. b	36. a	37. b	38. b	39. b	40. a
41. b	42. a	43. a	44. b	45. a	46. b	47. a	48. d	49. c	50. a
51. a	52. b	53. a	54. b	55. b	56. b	57. c	58. b	59. b	60. a
61. a	62. a	63. b	64. d	65. d	66. d	67. d	68. a	69. c	70. d
71. b	72. d	73. b	74. a	75. b	76. c	77. d	78. d	79. c	80. b
81. a	82. b	83. a	84. d	85. b	86. a	87. b	88. c	89. b	90. b
91. d	92. c	93. b	94. b	95. a	96. b	97. a	98. c	99. b	100. a
101. a	102. a	103. d	104. c	105. c	106. a	107. a	108. b	109. b	110. d
111. c	112. a	113. a	114. d	115. d	116. b	117. b	118. a	119. a	120. b
121. d	122. c	123. b	124. b	125. b	126. a	127. c	128. d	129. d	130. b
131. d	132. a	133. a	134. d	135. d	136. d	137. b	138. a	139. b	140. d
141. d	142. c	143. c	144. d	145. c	146. b	147. b	148. c	149. b	150. c
151. a	152. a	153. d	154. c	155. c	156. c	157. d	158. c	159. d	160. a
161. c	162. d	163. b	164. d	165. b	166. c	167. c	168. c	169. a	170. c

6

Overall Food Technology

1. Spirulina is a ---------
   (a) Edible fungus (b) Biofertilizer (c) Biopesticide (d) Single cell protein
2. Emulsifiers are used in commercially baked products to------------
   (a) Provide moistness (b) Improve flavor
   (c) Increase shelf-life (d) Reduce calories
3. The Healthy Mediterranean Diet Pyramid limits -----------
   (a) Physical activity (b) Whole grains (c) Meat and eggs (d) Beans and nuts
4. Good sources of protein in the diet are--------------
   (a) Fats, oils, butter, and margarine
   (b) Meats, fish, legumes, nuts, dairy products, and eggs
   (c) Deep green and orange vegetables, and citrus fruits
   (d) Green pepper, cantaloupe, citrus fruits, and broccoli
5. Which of the following vitamins is required for DNA synthesis?
   (a) Folate (b) Riboflavin (c) Pantothenic acid (d) Pyridoxine
6. In pre-cooling, heat is removed by------------
   (a) Convection (b) Conduction (c) Radiation (d) None of the above
7. The yoghurt is made from------------
   (a) *Lactobacillus bulgaricus* (b) *Streptococcus thermophilus*
   (c) *S. cremoris* (d) Mixed culture of a and b
8. Foods high in added sugars and saturated fats are-----------
   (a) High in fiber (b) Less nutrient dense
   (c) Low in caloric content (d) Not allowed in the diet
9. Cloudiness of brine of glass-packed olives may be caused by---------------
   (a) The lactics if fermentation of residual sugar resumes
   (b) Salt-tolerant bacteria
   (c) Film yeasts or moulds if air is available
   (d) All of the above

10. The chief spoilage organisms on smoked fish are-------------
(a) Molds (b) Bacteria (c) Both a and b (d) Fungi
11. Which country is the world's second largest exporter of quality wines?
(a) South Africa (b) Italy (c) France (d) Australia
12. Trans fats lowers------------
(a) LDL cholesterol (b) Shelf-life (c) Heart disease (d) HDL cholesterol
13. An obligate halophile requires high-----------
(a) pH (b) Temperature (c) Salt (d) Pressure
14. Which of the following substances are commonly used in a filter?
(a) Charcoal (b) Sand (c) Both a and b (d) Aluminum chloride
15. Which of the following is not a differential pressure flow meter?
(a) Flow nozzle (b) Rotameter (c) Venturimeter (d) Orificemeter
16. A supply chain is a sequence of firms that perform activities required------------
(a) To find products that are similar
(b) To facilitate wholesalers inventory selections
(c) To create synergy in their training programs
(d) To create and deliver goods to consumers
17. Milk fermentation to produce cheese is done initially by inoculating with----------
(a) *Saccharomyces cerevisiae*
(b) *Streptococcus lactis and Lactobacillus spp.*
(c) *Acetobacter and Gluconobacter*
(d) *Lactobacillus bulgaricus and Streptococcus thermophilus*
18. Agricultural produce (Grading and Marking Act 1937) act defines--------------
(a) Purchase (b) Sales (c) Storage (d) Quality by grade
19. Polyunsaturated fatty acids are found in------------
(a) Animal products like beef and lamb (b) Olive oil
(c) Fish oil (d) None of the above
20. Both temporary and permanent hardness of water can be removed on boiling water with-------
(a) Calcium hydroxide (b) Sodium carbonate
(c) Calcium oxide (d) Calcium carbonate
21. Personnel working in the market research group is responsible for the job of------------
(a) Product evaluation (b) Cost estimation
(c) Equipment selection (d) Equipment design
22. The juice that is extracted from the grapes for wine making is called------------
(a) Clear run juice (b) Solution (c) Must (d) Wort

23. Identify which of the following terms refers to the overall three dimensional shape of a protein.
(a) Primary structure (b) Secondary structure (c) Tertiary structure (d) Quaternary structure

24. In batch fermentation--------------
(a) Substrates are added to the system all at once and runs until product is harvested
(b) Nutrients are continuously fed into reactor and the product is siphoned off during the run
(c) New batches of microorganisms are screened for increased yield
(d) Small-scale production is used to synthesize product

25. Examination of the presence / absence of organisms refers to--------------
(a) Incubation of a food suspension in an enrichment medium followed by inoculation on to a suitable selective medium
(b) Incubation of a food suspension in an enrichment medium followed by inoculation on to a non-selective medium
(c) Incubation of a food suspension in an enrichment medium
(d) None of the above

26. NaCl can act as----------------
(a) Antagonist at optimal concentrations
(b) Synergistically if added in excess of optimum level
(c) Both a and b
(d) None of the above

27. An excessively high temperature in the fermentation of sauerkraut may inhibit the growth of--------
(a) *Leuconostoc* (b) *Pediococcus cerevisiae*
(c) Both a and b (d) None of the above

28. Which of the following statements is true for a refractive index detector in HPLC?
(a) It is more sensitive than a UV detector (b) It can only be used for isocratic elutions
(c) It does not respond to many solutes (d) None of the above

29. Sample retention in the column is measured by-----------
(a) Retention time (b) Retention factor (c) Retention index (d) All of the above

30. Heat in BTU necessary to increase the temperature of 1 lb of gas and its accompanying vapour by 1°F is called the-----------
(a) Latent heat (b) Humid heat (c) Specific heat (d) Sensible heat

31. A compound synthesized by bacteria or fungi that destroys or inhibits the growth of other microbes is a/an---------------
(a) Synthetic drug (b) Antibiotic (c) Antimicrobial drug (d) Competitive inhibitor

32. Heat is transferred by means of three different processes, which one of the following does not belong to the list?
(a) Conduction (b) Convection (c) Radiation (d) None of the above

33. An adequate amount of tyrosine can reduce the requirement of phenyl alanine in the diet, this effect is said to be------------
(a) Mutual supplementation of nutrients (b) Sparing effect
(c) Specific dynamic action (d) None of the above

34. The GC trace obtained after an experiment is called a-----------
(a) Chromatophore (b) Chromatogram (c) Chromatograph (d) Graph

35. Knowing the Glycemic Index is good for-----------
(a) Diabetics (b) People who are overweight or obese
(c) Athletes (d) All of the above

36. Chemo autotrophs can survive on----------alone.
(a) Minerals (b) CO_2 (c) Minerals and CO_2 (d) Methane

37. FSSAI regulates and monitors the safety of food to ensure-----------
(a) Manufacture and processing of food products
(b) Distribution, sale and import of food products
(c) Prices and taxes of food products
(d) Both a and b

38. ------------is the nutrient that plays a vital role in temperature regulation and digestion.
(a) Fats (b) Fiber (c) Protein (d) Water

39. Compositional analysis of---------- is done using mass spectrometer.
(a) An isotope (b) An alloy (c) Natural gas (d) A solid

40. Proteins can be visualized directly in gels by-----------
(a) Staining them with the dye (b) Using electron microscope only
(c) Measuring their molecular weight (d) None of the above

41. A butter from which the water and milk solids are removed is called-----------
(a) Whipped butter (b) Clarified butter (c) Salted butter (d) Margarine

42. Which nutrient if not used is converted to fat and stored around the body?
(a) Fat (b) Carbohydrate (c) Protein (d) Vitamins

43. Regulation of most food additives is the responsibility of the--------------
(a) Centers for Disease Control and Prevention
(b) United States Department of Agriculture
(c) Food and Drug Administration
(d) Environmental Protection Agency

44. Green tea helps to reduce ------------
(a) High BP (b) Obesity (c) High cholesterol (d) All of the above

45. Thin layer chromatography is-------------
(a) Partition chromatography (b) Electrical mobility of ionic species
(c) Adsorption chromatography (d) None of the above

46. Which fatty acids are clear liquids at cold temperatures?
(a) Saturated fats (b) Monounsaturated oils
(c) Tropical oils (d) Polyunsaturated oils

47. The most concentrated source of Kcals is------------
(a) Starch (b) Protein (c) Alcohol (d) Sugar

48. Drying operation under vacuum is carried out to------------
(a) Dry those materials which have very high unbound moisture content
(b) Reduce drying temperature
(c) Increase drying temperature
(d) Dry materials having high bound moisture content

49. As per FSSA, 2006, after receiving the sample the food analyst should submit the analysis report within --------------days.
(a) 13 (b) 14 (c) 15 (d) 16

50. Lipoic acid is a co-factor of the------------
(a) Pyruvic dehydrogenase (b) α-ketoglutaric dehydrogenase
(c) Both a and b (d) Di-hydroorotate dehydrogenase

51. Which of the following statements about the use of mass spectrometry in protein investigation is correct?
(a) Mass spectrometry involves ionized molecules in the liquid phase
(b) Mass spectrometry is used for analysing the three-dimensional shape of peptides
(c) Ionic fragments are separated according to their mass-to-charge ratio in Mass spectrometry
(d) Mass spectrometry involves the separation of ionic fragments on a gel

52. Toxic metal contamination can occur through-----------
(a) Serving foods in lead glazed ceramics
(b) Shellfish harvested from waters with high levels of mercury or lead
(c) Cooking acidic foods, such as tomatoes, in an unlined copper pan
(d) All of the above

53. The presence of coliform bacteria in a pasteurized milk sample is an indication of----------
(a) Improper storage (b) High quality milk
(c) Poor sanitation (d) None of the above

54. Detergents are-------------
(a) High-level germicides (b) Low-level germicides
(c) Excellent antiseptics (d) Used in disinfecting surgical instruments

55. Which of the following is an approved sterilant?
(a) Chlorhexidine (b) Betadyne (c) Ethylene oxide (d) Ethyl alcohol

56. During the freezing process ------------- can damage cell walls leading to changes in product texture and quality.
(a) Sugar crystals (b) Salt crystals (c) Carbon dioxide (d) Ice crystals

57. Which of the following is a heterofermentative micro-organism?
(a) *Leuconostoc mesenteroides* (b) *Lactobacillus delbrueckii*
(c) *Lactobacillus bulgaricus* (d) *Lactobacillus pentose*
58. The red or pink color of the fish is generally caused from the growth of---------------
(a) *Sarcina* (b) *Micrococcus or Bacillus species*
(c) *Molds or yeasts* (d) All of the above
59. Oil and water separate when mixed together due to the ------------ portions of fatty acids in oil.
(a) Hydraulic (b) Hydrophilic (c) Hydrophobic (d) Hydroscopic
60. An extruder is a equipment commonly used to form------------
(a) Meat patties (b) Crackers (c) Tortillas (d) Pasta
61. -----------are the major parts of a wheat kernel.
(a) Germ, bran, endosperm and husk (b) Germ, bran and endosperm
(c) Fiber, germ bran and husk (d) Oil, bran, endosperm and hull
62. To produce sauerkraut, cabbage undergoes a ------------- process.
(a) Fermentation (b) Drying (c) Freezing (d) Cold sterilization
63. Essential amino acids--------------
(a) Are produced in our body through biochemical pathways
(b) Are only acquired through dietary intake
(c) Are not necessary for survival
(d) Are available in protein free foods
64. Which of the following is not used as stationary phase in a GC column?
(a) Polysiloxanes (b) Silica (c) Cyclodextrins (d) None of the above
65. To determine the amount of protein in the body, an individual's -----------can be measure(d)
(a) Nitrogen levels (b) Cholesterol levels (c) Oxygen levels (d) Carbon levels
66. Gel chromatography separates molecules according to their --------------
(a) Molecular size (b) Polarity (c) Solubility (d) Matrix
67. Biogas is -----------
(a) Methane rich fuel (b) Ecofriendly and pollution free source
(c) Propane rich fuel (d) Both a and c
68. High value of BOD (Biochemical Oxygen Deman(d) shows ---------
(a) Water is normal (b) Water is highly polluted
(c) Water is less polluted (d) None of the above
69. What is the large-scale study of cellular proteins called?
(a) Proteomics (b) Genomics (c) Metabolomics (d) Bioinformatics
70. Which vitamin enhances iron absorption?
(a) Thiamin (b) Ascorbic acid (c) Riboflavin (d) Folate

71. Acidophilus milk which is used for its therapeutic properties in intestinal disorders requires a pure culture of-------------
(a) *Lactobacillus bulgaricus* (b) *Streptococcus thermophilics*
(c) *Lactobacillus acidophilus* (d) None of the above

72. Which of the following(s) is/are the advantage of HPLC over traditional LPLC (low-pressure liquid chromatography)?
(a) Greater sensitivity and reusable columns (b) Ideal for ionic species and large molecules
(c) Sample recovery (d) All of the above

73. Relative humidity is the ratio of the-------------
(a) Actual humidity to saturation humidity
(b) Partial pressure of the vapour to the vapour pressure of the liquid at gas temperature
(c) Partial pressure of the vapour to the vapour pressure of the liquid at room temperature
(d) None of the above

74. Break-even point is the point of intersection of----------
(a) Fixed cost and total cost (b) Total cost and sales revenue
(c) Fixed cost and sales revenue (d) None of the above

75. The maximum permissible limit (BIS) of turbidity in drinking water is------------
(a) 5 NTU (b) 10 NTU (c) 15 NTU (d) 20 NTU

76. In a fluidized bed reactor---------------
(a) Hot spots are formed (b) Temperature is more or less uniform
(c) Temperature gradients are very high (d) Segregation of the solids occurs

77. The common methods used for disinfection of waste water treatment plants are-----------
(a) Chlorination (b) UV light (c) Both a and b (d) Phenolic solvent

78. Which of the fluid forces are not considered in the Reynold's equation of flow?
(a) Compressibility forces (b) Viscous forces
(c) Turbulent forces (d) Pressure forces

79. Which of the following is likely to improve the LDL/HDL ratio, in terms of the prevention of atherosclerosis?
(a) Monounsaturated fats (b) Polyunsaturated fats
(c) Fish oils (d) Omega- 4 fats

80. Riboflavin is a by-product of -----------------fermentation carried out by *Clostridium* spp.
(a) Citric acid (b) Acetone butanol (c) Streptomycin (d) Ethanol

81. Who showed that *Sacchaaromyces cerevisiae* causes fermentation forming products such as beer and buttermilk?
(a) Louis Pasteur (b) Alexander Fleming
(c) Selman Waksman (d) Schatz

82. In a gel filtration column---------------
(a) Smaller proteins enter the beads more readily
(b) Large proteins elute first
(c) Both a and b
(d) Large proteins enter the beads more readily

83. Rennet is used in ----------
(a) Bread making (b) Fermentation (c) Cheese making (d) Antibiotics synthesis

84. To decrease LDL-cholesterol, researchers suggest eating foods high in ------------
(a) Carbohydrate (b) Saturated fat (c) Fat (d) Soluble fiber

85. What metal ion is specifically bound by vitamin B12?
(a) Cobalt (b) Copper (c) Zinc (d) Iron

86. Which one of the following is added to wines for making fortified wine?
(a) Gin (b) Brandy (c) Whisky (d) Rum

87. Which one of the following causes astringency in red wine?
(a) Tannin (b) Honey (c) Lemon (d) Vinegar

88. In gas chromatography, the basis for separation of the components is the difference in----------
(a) Partition coefficients (b) Conductivity
(c) Molecular weight (d) Molarity

89. Characteristic feature of any form of chromatography is the -----------
(a) Use of molecules that are soluble in water
(b) Use of an inert carrier gas
(c) Calculation of an Rf value for the molecules separated
(d) Use of a mobile and a stationary phase

90. Bioreactor is ---------
(a) Hybridoma (b) Culture containing radioactive isotopes
(c) Culture for synthesis of new chemicals (d) Fermentation tank

91. The toxin of which of the following is used as a biological insecticide?
(a) *Pseudomonas syringae* (b) *Botryococcus braunii*
(c) *Bacillus thuringiensis* (d) *Aspergillus oryzae*

92. Pascal law is not applicable for a -------------- fluid.
(a) Uniformly moving (b) Static
(c) Accelerating frictionless (d) None of the above

93. Fermented foods are usually an example of which of the following?
(a) Highly perishable food (b) Perishable food
(c) Semi perishable food (d) Nonperishable food

94. Sedimentation is a physical process used in waste water treatment to------------
   (a) Remove particles that are less dense than water
   (b) Remove particles that are more dense than water
   (c) Remove the pertinacious material from the water
   (d) None of the above
95. Caffeine is found in which of the following?
   (a) Tea leaves (b) Coffee seeds (c) Kola seeds (d) All of the above
96. Triglycerides are so named because they are formed by a reaction between three fatty acid molecules and one---------------
   (a) Glucose (b) Glycerol (c) Amino acid (d) Lipid
97. Enzymes are one type of protein, they function to-----------------
   (a) Form hormones (b) Repair tissue (c) Build strength (d) Speed reactions
98. In FSSAI, out of 22 members------------- shall be women members.
   (a) 1/5 (b) 1/4 (c) 1/3 (d) 1/2
99. ------------- pumps are axial flow pumps.
   (a) Propeller (b) Turbine (c) Diffuser (d) None of the above
100. Which of the following is classified as "steroid?"
   (a) Phospholipid (b) Glycerol (c) Wax (d) Cholesterol
101. Which of the following is biodegradable plastic?
   (a) Poly propylene (b) LDPE (c) HDPE (d) Polyhydroxybutyrate
102. Which one of the technologies listed below is a valuable method for mass-producing drugs and other useful proteins?
   (a) Recombinant DNA technology (b) Transgenic technology
   (c) Biotechnology (d) Gene targeting
103. Enzyme immobilization is -----------
   (a) Conversion of an active enzyme into inactive form
   (b) Providing enzyme with protective covering
   (c) Changing a soluble enzyme into insoluble state
   (d) Changing pH so that enzyme is not able to carry out its function
104. Biogas is produced by -------
   (a) Aeobic breakdown of biomass (b) Anaerobic breakdown of biomass
   (c) With the help of methanogenic bacteria (d) Both b and c
105. A risk factor for heart disease includes--------------
   (a) High HDL-cholesterol level (b) Active lifestyle
   (c) Healthy body weight (d) Type 2 diabetes

106. Permanent hardness of water is caused by the presence of-----------
(a) Bicarbonates of calcium and magnesium
(b) Carbonates of sodium and potassium
(c) Chlorides and sulfates of calcium and magnesium
(d) Phosphates of sodium and potassium

107. In SDS-PAGE, the protein sample is first--------------
(a) Treated with a reducing agent and then with anionic detergent followed by fractionation by electrophoresis
(b) Fractionated by electrophoresis then treated with an oxidizing agent followed by anionic detergent
(c) Treated with a oxidizing agent and then with anionic detergent followed by fractionation by electrophoresis
(d) None of the above

108. Vinegar is obtained from molasses with the help of ------------
(a) Rhizopus (b) Acetobacter (c) Yeast (d) Both b and c

109. Which of the following is not essential to carry out the polymerase chain reaction?
(a) Primers (b) DNA polymerase (c) Gel electrophoresis (d) High temperature

110. Any process that destroys the non-spore-forming contaminants on inanimate objects is-------
(a) Antisepsis (b) Disinfection (c) Sterilization (d) De germination

111. The functions of Designated Officer is---------
(a) To issue or cancel the license of food business operator
(b) To prohibit the sale of article which is in contravention of the provision in act
(c) To receive the report of articles
(d) All of the above

112. In reversed-phase HPLC-------------
(a) A hydrophobic stationary phase is combined with a polar mobile phase
(b) A hydrophobic stationary phase is combined with a non-polar mobile phase
(c) A hydrophilic stationary phase is combined with a polar mobile phase
(d) A hydrophilic stationary phase is combined with a non-polar mobile phase

113. The micro-organism which is present in both fresh and frozen juices is----------
(a) *E. coli* (b) *Entereobacter aerogenes*
(c) *P. chrysogenum* (d) None of the above

114. Molds causing spoilage of eggs include species of---------------
(a) *Cladosporium* (b) *Mucor* (c) *Thamnidium* (d) All of the above

115. What are the intrinsic factors for the microbial growth?
(a) pH (b) Oxidation-Reduction Potential
(c) Moisture (d) All of the above

116. The enzymatic browning is due to the oxidation of phenols to -----------

(a) Glucocinolate (b) Tannins (c) Prolamins (d) Orthoquinones

117. The partial or complete disorganization of a protein's three-dimensional shape is called ------

(a) Oxidation (b) Reduction (c) Hydrolysis (d) Denaturation

118. ------------- is rich source of omega-3 fatty acid.

(a) Olive oil (b) Canola oil (c) Salmon (d) Cod-liver oil

119. Mass transfer rate between two fluid phases does not necessarily depend on the ----------- of the two phases.

(a) Chemical properties (b) Degree of turbulence

(c) Interfacial area (d) Physical properties

120. With increase in temperature, the vapor pressure of liquids----------

(a) Increases linearly (b) Decreases

(c) Remains constant (d) Increases

121. A major organic acid produced by a microbial process used in foods is--------------

(a) Sulfuric acid (b) Citric acid (c) Oxalic acid (d) Uric acid

122. If acidic foods (such as tomatoes) are added to milk, then-------------

(a) Fat coagulates (b) Fat content increases

(c) Casein coagulates (d) Whey coagulates

123. Which of the following physical method is used as germicidal in modern time for the treatment of drinking water?

(a) Chlorination (b) Treating with potassium permagnate

(c) UV radiation (d) Treating with bleaching powder

124. High blood levels of ------- substance may indicate there is inflammation in the arterial walls.

(a) HDL-cholesterol (b) Estrogen

(c) C-reactive protein (d) Insulin

125. Which one of the following is incorrect regarding labeling of food package?

(a) Product identification

(b) Name and address of the manufacturer, packer, or distributor

(c) Net contents or net weight

(d) Sources of food ingredients

126. Which is NOT a function of lipid?

(a) Enhances food flavor (b) Primary storage of energy in the body

(c) Component of cell membranes (d) Primary source of energy for the brain

127. Antioxidants perform all of the following except-------------

(a) Prevent protein degradation (b) Preserve color

(c) Minimize rancidity (d) Preserve flavor

128. Microbe(s) that is/are the target(s) of pasteurization include------------
(a) *Clostridium botulinum* (b) *Mycobacterium species*
(c) *Salmonella species* (d) Both b and c

129. A chemical linked to long-term effect such as cancer, sterility and birth defects could cause which of the following-------------
(a) Chronic toxicity (b) Acute toxicity (c) Defect action levels (d) Total adverse response

130. A toxin commonly found in corn and peanuts is------------
(a) Solanine (b) Protease (c) Goitrogens (d) Aflatoxins

131. Which of the following statements is FALSE?
(a) LDLs transport cholesterol to the arterial wall
(b) Smoking is a risk factor of heart disease
(c) HDLs will eventually increase with weight loss
(d) None of the above

132. A food that can be stored at room temperature for a prolonged or indefinite time period with minimal quality deterioration is said to be------------
(a) Room stable (b) Shelf superior (c) Shelf stable (d) Room superior

133. Chemically leavened dough uses ------------ as a leavening agent.
(a) Air (b) Baking powder (c) Yeast (d) Steam

134. Milk undergoes a process called------------ that is intended to break down fat globules so they are smaller and more uniform in size.
(a) Homogenization (b) Pasteurization (c) Encapsulation (d) Emulsification

135. As fruits mature, the specific gravity will------------
(a) Increases (b) Decreases (c) Remains constant (d) None of the above

136. -------------is a system for monitoring food production for compliance with health, safety and product standards.
(a) Research and development (b) HACCP
(c) Quality assurance (d) Government inspection

137. Mostly dry fruit are rich in------------
(a) Protein (b) Carbohydrates (c) Fats (d) Vitamins

138. When consumption of food results in death of the customer, the provision of penalty under FSSA 2006 is---------------
(a) Imprisonment not less than 7 years (b) Can be extended for life imprisonment
(c) Fine not less than 10 lakh (d) All of the above

139. Enzyme responsible for converting protopectin into pectin is ---------
(a) PME (b) Proto-pectinase (c) Poly galacturonase (d) Pectinase

140. Which of the following is/are non-climacteric type of fruits?
(a) Pineapple (b) Litchi (c) Grape (d) All of the above

141. Vacreation refers to ---------

(a) Vacuum packaging (b) Vacuum creation
(c) Vacuum sterilization (d) Vacuum pasteurization

142. What is AFD?

(a) Accelerated Fish Drying (b) Accelerated Force Drying
(c) Accelerated Freeze Drying (d) Accelerated Frequencing Drying

143. For distant marketing, tomato fruits are harvested at----------

(a) Immatured green stage (b) Matured green stage
(c) Turning stage (d) Red ripe stage

144. ISO 22000 standards are associated with------------

(a) Safety and quality (b) Quality
(c) Safety (d) None of the above

145. ISO 9002 requires -------------

(a) Getting production and inspection/installation methods certified
(b) Frequent vendor visits
(c) Constantly holding meetings with customers
(d) Getting final inspection methods shipshape

146. Vitamins are essential because the organism-------------

(a) Can't synthesize these compounds at all
(b) Can synthesize these compounds partially
(c) Can't synthesize these compounds in the adequate amounts
(d) None of the above

147. The vitamin niacin is part of which molecule?

(a) Ferredoxin (b) Pyridoxal phosphate
(c) Pyrophosphate (d) NAD+

148. The most prominent role that tocopherol has in in-vitro systems is as a strong-------------

(a) Antioxidants (b) Reducing agent (c) Oxidizing agent (d) All of the above

149. SWMA stands for-----------

(a) Standards of Weights and Monitoring Act, 1976
(b) Standards of Water and Monitoring Act, 1976
(c) Standards of Water and Management Act, 1976
(d) Standards of Weights and Measure Act, 1976

150. Which of the following packages is an example of aseptic packaging?

(a) Tetra pak (b) Paper bag (c) Milk carton (d) Plastic bread bag

151. An example of an insoluble fiber is--------------

(a) Pectin (b) Cellulose (c) Gums (d) All of the above

152. What is the process of heating liquids or food to destroy food borne pathogens?
(a) Modified atmosphere packaging (b) Food preservation
(c) Pasteurization (d) High-pressure processing

153. Which of the following can cause chemical contamination of food from food handlers?
(a) Hair (b) Dust (c) Live insects (d) Perfume

154. Cross-contamination can be reduced by-------------
(a) Personal cleanliness and hygiene
(b) Labeling all food items
(c) Reducing the time that food is in the danger zone
(d) All of the above

155. A brown color in milk may result from-------------
(a) *Pseudomonas putrefaciens*
(b) The enzymatic oxidation of tyrosine by *P. fluorescens*
(c) Both a and b
(d) *S. marcescens*

156. The term empty calories refers to foods that---------------
(a) Are very high in fats
(b) Are low in both calories and nutrients
(c) Are high in calories but low in nutrients
(d) Provide nutrients above and beyond the recommended daily allowances

157. Which vegetarian diet excludes all animal products?
(a) Vegan (b) Ovovegetarian (c) Lactovegetarian (d) Ovolactovegetarian

158. ISO 22000 standards are associated with------------
(a) Safety and quality (b) Quality
(c) Safety (d) None of the above

159. Oil and water normally separate because they are---------------
(a) Emulsified (b) Immiscible
(c) Colloidal dispersion (d) None of the above

160. -------------is an illness caused by consuming a food that contains a harmful metabolite from microorganisms.
(a) Food borne infection (b) Baceriocide
(c) Bacteriostat (d) Food borne intoxication

161. Fruits and vegetables discolor when bruised or cut due to-------------
(a) Caramelization (b) Sulfiting (c) Dehydration (d) Enzymatic browning

162. A small cardboard ticket consisting coded information and is usually attached to products is usually classified as------------
(a) Tags (b) Coded cardboard (c) Decoded paper (d) Digital cardboard

163. One of the functions of sodium nitrite in meat products is to-----------

(a) Inhibit mold growth
(b) Inhibit growth of *Clostridium botulinum* in vacuum packaged cured meats
(c) Minimize purge in vacuum packaged meats
(d) Reduce color fading in aerobically packaged cured meats

164. Which of the following is NOT a protein?

(a) Maltose (b) Hemoglobin (c) Insulin (d) Collagen

165. Set of parallel printed lines with different thickness with black and white characters is called---------------

(a) White code (b) Magnetic code (c) Colored code (d) Bar code

166. Microfiltration membrane is not made of ------------

(a) Polysulfone (b) Cellose acetate (c) Ceramic (d) Poly propylene

167. The three-carbon "backbone" found in all triglycerides is called------------

(a) Glycerol (b) Fatty acid (c) Acetic acid (d) Stearic acid

168. A significant health hazard from obesity is------------

(a) Ulcers (b) Diverticulosis (c) Type 2 diabetes (d) Emphysema

169. A special beer yeast of the bottom type is--------------

(a) *S. cerevisiae* (b) *S. carlsbergensis* (c) *S. thermophillus* (d) None of the above

170. When the fermentation of fructose results in the bitter product mannitol, the fermentation is sometime termed as----------------

(a) Mannkic (b) Amertume (c) Mannitic (d) None of the above

171. The best foods to lose weight should contain some -------------

(a) Protein (b) Complex carbohydrates
(c) Fat (d) None of the above

172. Which protein source is most likely to lower blood levels of cholesterol?

(a) Chicken (b) Pork (c) Fish (d) Soy

173. Low thermal conductivity of heat insulating materials is due to its----------

(a) Dense structure (b) High proportion of air space
(c) High specific heat (d) None of the above

174. Chromatography is based on the -----------------

(a) Different rate of movement of the solute in a column
(b) Separation of one solute from other constituents by being captured on the adsorbent
(c) Different rate of movement of the solvent in the column
(d) None of the above

175. Once infected with *Escherichia coli* O157:H7, how does one prevent its spread?

(a) Keep food away from animals (b) A hospital stay, ingesting clean, clear fluids
(c) Boil water for at least 10 minutes (d) Good hygiene

176. A leading cause of high bacterial counts in raw milk is------------
(a) Younger cows (b) BST (c) Poor sanitation (d) Older cows

177. The most common form of quality control includes--------------
(a) Planning (b) Organizing (c) Directing (d) Inspection

178. Zeolite softening process removes------------
(a) Only temporary hardness of water
(b) Only permanent hardness of water
(c) Both temporary and permanent hardness of water
(d) The dissolved gases in permanent hard water

179. In TLC the strength with which an organic compound binds to an adsorbent depends on------
(a) Ion- dipole (b) Hydrogen bonding (c) Dipole- dipole (d) All of the above

180. Which organic solvent is used for yeast cell lysis?
(a) Acetonitrile (b) Carbinol (c) Dichloromethane (d) Toluene

181. Out of these quality standards which are mandatory standards?
(a) Legal standards (b) Company standards (c) Industry standards (d) Grade standards

182. The neurotransmitter serotonin is synthesized from which amino acid?
(a) Phenylalanine (b) Tyrosine (c) Cysteine (d) Tryptophan

183. Butter is made by agitating cream to form a------------- emulsion.
(a) Water-in-oil (b) Gas-in-liquid (c) Oil-in-water (d) Gas-in-solid

184. The water activity of pure water is-------------
(a) 0.100 (b) 1.000 (c) 10.00 (d) 100.0

185. The sugar ------------ is sweeter than sucrose.
(a) Fructose (b) Lactose (c) Glucose (d) Maltose

186. To control crystal size when making candy, an interfering agent such as------------ is added.
(a) Salt (b) Sugar (c) Water (d) Cream of tartar

187. TDT is ----------------
(a) Thermal dealy time (b) Thermal divide time
(c) Thermal death temperature (d) Thermal death time

188. In running food industry identifying and solving the problems is known as ------------
(a) Monitoring quality assurance (b) Troubleshooting
(c) Implementing HACCP (d) Pilot testing

189. Kg of liquid evaporated per hour in an evaporator is defined as its-------------
(a) Capacity (b) Economy (c) Steam load (d) None of the above

190. What can occur in atherosclerosis?
(a) Occlusion or blocking of the arteries (b) Heart attack
(c) Cholesterol deposition (d) All of the above

191. What is a benefit of eating plant proteins instead of animal proteins?

(a) High in fiber (b) Low in cholesterol (c) More energy (d) Both a and b

192. --------- is the separation in which mobile phase composition remains constant throughout the procedure.

(a) Isostatic (b) Isocratic (c) Gradient elution (d) Both a and b

193. What element forms the skeleton of organic molecules?

(a) Hydrogen atoms (b) Phosphate atoms (c) Carbon atoms (d) Water molecules

194. What happens in a dehydration reaction?

(a) Molecules are broken apart

(b) Monomers arc bonded together and a water molecule is released

(c) Atoms are joined

(d) It depends on what molecule it is

195. Which enzymes exist in two or more forms have same function, but differ physically and chemically?

(a) Apoenzymes (b) Multienzymes (c) Exoenzymes (d) Isoenzymes

196. Which of the following techniques are used by FDA in determining the manner in which industry accepts its responsibilities?

(a) Establishment inspection

(b) Sample collection and analysis of product in process

(c) Surveillance intended to identify problems

(d) All of the above

197. Gliadin is soluble in -----------

(a) Salt (b) Alcohol

(c) Dilute acid and alkali (d) Sugar

198. Anatomy of barcode consists of--------------

(a) Number of system character (b) Manufacture and product code

(c) Guard bars and check digit (d) All of the above

199. Self cooling packages contains--------------

(a) Ammonium bicarbonate (b) Ammonium chloride

(c) Sodium hydroxide (d) Calcium chloride

200. Where do substrates attach to an enzyme?

(a) Peptide bond (b) Ring binding site (c) Active site (d) Enzymatic site

201. What is extraction in paprika used for?

(a) Obtain proteins from the end product (b) Desolventization of final product

(c) Extraction of its natural color (d) All of the above

202. Protein synthesis is regulated principally by-----------

(a) Glucocorticoids (b) Adrenocorticotropic hormone

(c) Insulin (d) None of the above

203. The neurotransmitters adrenaline, noradrenalin and dopamine are synthesized from which amino acid?
(a) Glycine (b) Arginine (c) Tyrosine (d) Glutamine

204. Which one of the following neurotransmitters is produced from the amino acid tryptophan?
(a) Serotonin (b) Histamine (c) Acetylcholine (d) Nitric oxide

205. If an essential amino acid that is needed by cells to make a particular protein is unavailable in the body pool, the cells will----------
(a) Omit that amino acid from the protein (b) Substitute a similar amino acid
(c) Break down body protein to obtain it (d) Synthesize the amino acid

206. The use of nitrogen measurement to assess protein balance is based on the assumption that proteins contain how much nitrogen?
(a) 8% (b) 16% (c) 24% (d) 32%

207. The practice of mixing different classes of dietary proteins so that the deficits in one are balanced by the surplus in the other is called-------------
(a) Protein quality (b) Protein digestibility
(c) Protein complementation (d) Nitrogen balance

208. Which one of the following definitions represents the net protein utilization (NPU) value of a protein?
(a) The proportion of consumed nitrogen that is absorbed
(b) The proportion of absorbed nitrogen that is retained in the body
(c) The proportion of consumed nitrogen that is retained
(d) None of the above

209. Which of the following types of molecules is found in genetic material?
(a) Cellulose (b) Enzymes (c) Lipids (d) Nucleic acids

210. What is the water-soluble vitamin exception which humans can store in large amounts?
(a) Niacin (b) Cobalamin (c) Pyridoxine (d) Thiamin

211. Which of the following can cause food to be contaminated because of physical hazards from food handlers?
(a) Jewellery (b) Dust
(c) Rodent droppings (d) Incorrectly diluted chemicals

212. The Hepatitis A virus is-------------
(a) Often caused by contaminated shellfish (b) Carried by humans
(c) Transmitted by poor personal hygiene (d) All of the above

213. Which of the following pathogens is considered a Category A biological agent?
(a) *Escherichia coli* O157:H7 (b) *Clostridium botulinum*
(c) *Vibrio cholerae* (d) *Cryptosporidium parvum*

214. --------------consist of carbohydrates, fats, and proteins.

(a) Vitamins (b) Micronutrients (c) Phytochemicals (d) Macronutrients

215. The primary type of starter culture used in cheddar cheese manufacture is---------

(a) Psychotropic bacteria (b) Lactic acid bacteria

(c) Mold (d) None of the above

216. A successful TQM program incorporates all of the following except -----------

(a) Continuous improvement (b) Employment involvement

(c) Centralized decision making authority (d) Benchmarking

217. Which of the following are found in nucleotides of DNA and RNA respectively?

(a) Deoxyribose and ribose (b) Proteins and enzymes

(c) Fats and oils (d) Sugars and starches

218. What is the function of ATP, adenosine triphosphate?

(a) Message carrier (b) Store and transport energy

(c) Make proteins (d) Breakdown sugars

219. The growth of black-pigmented *Bacillus nigrificans*, a causative agent of black pickle is favoured by---------------

(a) Presence of carbohydrate like glucose (b) Low level of available nitrogen

(c) Neutral or slightly alkaline brine (d) All of the above

220. Slippery pickles occur due to----------------

(a) Growth of encapsulated bacteria

(b) Broken scum of film yeasts on the surface of brine

(c) Addition of high amount of salt

(d) Both a and b

221. All of the following foods are high in carbohydrate except--------------

(a) Milk (b) Macaroni (c) Lentils (d) Margarine

222. Examples of digestive enzymes include---------------

(a) Amylase (b) Lactose (c) Maltose (d) All of the above

223. The standard plate count of milk reports the-------------

(a) Percent protein (b) Number of aerobic bacteria in a sample

(c) Number of anaerobic bacteria in a sample (d) Somatic cell count

224. Total Quality Management emphasizes ------------

(a) The responsibility of the Quality Control staff to identify and solve all quality-related problems

(b) A system where strong managers are the only decision makers

(c) A commitment to quality that goes beyond internal company issues to suppliers and customers

(d) A process where mostly statisticians get involved

225. Fermentation of grains usually results in production of --------------
(a) Carbon dioxide (b) Brine (c) Must (d) All of the above

226. Which nutrient is needed in tiny quantities to help in the body's chemical processes?
(a) Minerals (b) Carbohydrate (c) Protein (d) Vitamins

227. At constant temperature, the thermal conductivities of gases ------------ with rise in pressure.
(a) May increase or decrease; depends on the pressure
(b) Decrease
(c) Increase
(d) Remain unchanged

228. Which of the following foods is a reliable source of sodium?
(a) Apple (b) Skinless chicken breast
(c) 2% milk (d) Dill pickle

229. Which trace mineral's primary function is glucose transport?
(a) Selenium (b) Manganese (c) Chromium (d) Iodine

230. Ascorbic acid acts as a-------------
(a) Reducing agent (b) Oxidizing agent
(c) Oxidizing and reducing agent both (d) None of the above

231. Vitamin B_{12} (Cobalamin) is only synthesized by-----------
(a) Fishes (b) Microorganisms (c) Plants (d) Animals

232. Which of the following are reduced coenzymes?
(a) NADH and FADH2 (b) NAD+ and FAD
(c) ATP and GTP (d) Coenzyme A and ubiquinone

233. Lipoic acid exists in-----------
(a) Oxidized form (b) Reduced form (c) Both a and b (d) None of the above

234. Which vitamin may interact with the anti-coagulant warfarin and increase the risk of excessive bleeding?
(a) Vitamin A (b) Vitamin E (c) Vitamin B_6 (d) Vitamin B_{12}

235. Which one of the following foods does not contain retinol?
(a) Liver (b) Milk (c) Eggs (d) Carrots

236. Which one of the following statements is incorrect?
(a) Tocopherols have a saturated hydrocarbon chain, whereas tocotrienols are unsaturated
(b) Tocopheryl esters are more stable than free tocopherols
(c) α-, β-, γ- and δ-tocopherols differ from each other on the hydrocarbon chain
(d) Vitamin E acts as an antioxidant in the lipid environment of cells

237. The synthetic form of vitamin K is called------------
(a) Menaquinone (b) Menadione (c) Hydroquinone (d) Phylloquinone

238. Of the following food groups, which one is the richest dietary source of phylloquinone?
(a) Meats (b) Dairy products
(c) Green leafy vegetables (d) Citrus fruits

239. The blood-thinning agent warfarin prevents the recycling of------------
(a) Vitamin A (b) Vitamin D (c) Vitamin E (d) Vitamin K

240. Which one of these beverages is the richest source of riboflavin?
(a) Red wine (b) Blackcurrant juice (c) Whole milk (d) Black tea

241. Because of its metabolic role, vitamin B_6 requirements depend on intake of----------
(a) Carbohydrate (b) Fat (c) Protein (d) Energy

242. Biotin is required for the function of which one of the following enzymes?
(a) Acetyl CoA carboxylase (b) Transketolase
(c) Glycogen phosphorylase (d) Alanine aminotransferase

243. Which vitamin acts as a methyl (CH_3) group donor for the reconversion of homocysteine to methionine?
(a) Pantothenic acid (b) Niacin (c) Folate (d) Biotin

244. Synthetic vitamin B_{12} is called-------------
(a) Methylcobalamin (b) 5′-Deoxyadenosyl cobalamin
(c) Cyanocobalamin (d) None of the above

245. Which one of the following food groups would not be a good source of vitamin B_{12}?
(a) Dairy products (b) Green vegetables (c) Eggs (d) Fish

246. An example of a digestive hormone is--------------
(a) Lipase (b) Gastrin (c) Pepsin (d) Amylase

247. What compound of raw egg white causes a syndrome similar to vitamin B deficiency?
(a) Avidin (b) Betabindin (c) Ovalbumin (d) Albumin

248. A deficiency of Vitamin K results in a decreased level of------------
(a) Fibrin (b) Thrombin (c) Prothrombin (d) Fibrinogen

249. A fat-soluble vitamin that regulates blood clotting is--------------
(a) Vitamin A (b) Vitamin K (c) Vitamin C (d) Niacin

250. Which of the following is not a condition that bacteria would thrive and multiply?
(a) Dry (b) Nutrient-rich (c) Proper pH (d) Correct temperature

251. What is the process in which foods are placed in a shielded chamber and subjected to a radiant energy source?
(a) Irradiation (b) Modified atmosphere packaging
(c) Food preservation (d) High-pressure processing

252. Which of the following species of *Clostridium* is responsible for formation of dark green to black colours in cheese?
(a) *Clostridium tyrobutyricum* (b) *Clostridium sporogenes*
(c) *Clostridium herbarum* (d) None of the above

253. Sparkling wines are named because they contain------------

(a) Sulfur Dioxide (b) Carbon Dioxide (c) Hydrogen (d) Oxygen

254. A cluster of GI symptoms that occur after eating a particular food is called ----------

(a) Food allergy (b) GI disorders

(c) Food malabsorption (d) Food intolerance

255. According to the manufacturing-based definition of quality ----------

(a) Quality is the degree of excellence at an acceptable price and the control of variability at an acceptable cost

(b) Quality depends on how well the product fits patterns of consumer preferences

(c) Quality is the degree to which a specific product conforms to standards

(d) Even though quality cannot be defined, you know what it is

256. Pickles and sauerkraut share a common inoculums, which is -----------

(a) *Lactobacillus plantarum* (b) *Lactobacillis bulgaricus*

(c) *Lactobacillis acidophilus* (d) *Saccharomyces cerevisiae*

257. Secondary metabolites -----------

(a) Are essential to microbe function

(b) Are by-products of metabolism that are not important to microbe function

(c) Are products that require additional processing before they can be packaged

(d) Are harvested during the exponential phase of growth

258. The monosaccharides include------------

(a) Maltose, glucose and lactose (b) Fructose, galactose and sucrose

(c) Fructose, glucose and galactose (d) Fructose, glucose and lactose

259. Which nutrient helps to absorb vitamins A, D, E, and K?

(a) Fat (b) Carbohydrate (c) Protein (d) Vitamins

260. Which of the following lacks antioxidant properties?

(a) Vit E (b) Omega 3 fatty acids (c) Vit K (d) Calcium

261. Which of the following is a concern about GE foods?

(a) Safe animal feeds (b) Diminished level of toxins

(c) Desirable genetic modification (d) Introduction of allergens

262. What is the main type of micro-organism responsible for food poisoning?

(a) Bacteria (b) Mould (c) Virus (d) Parasite

263. Which of the following are allergens?

(a) Oilseeds (b) Fruits and vegetables

(c) Fish and fish products (d) Red meat

264. What will a Health Inspector normally do if he or she finds that food preparation premises are unsatisfactory and potentially dangerous to the public?
(a) Issue a court summons
(b) Demand the premises to be vacated
(c) Issue a Hygiene Emergency Prohibition Order
(d) Notify the police to make an arrest

265. A food that is contaminated with bacteria will------------
(a) Taste "off" (b) Emit a foul smell
(c) Appear discoloured (d) There are no outward warning signs

266. If plaster is found in a food item, what type of food hazard is this considered?
(a) Physical (b) Chemical (c) Biological (d) All of the above

267. What is the first thing that should be done before entering an area where food is being prepared?
(a) Go to the bathroom (b) Comb the hair
(c) Wash the hands (d) None of the above

268. Most food-borne illnesses go undiagnosed because-------------
(a) Symptoms may not appear for a week or more
(b) The symptoms are not serious enough to warrant a hospital visit
(c) Most victims die before treatment
(d) Symptoms usually appear during flu season

269. In order to survive and multiply, all bacteria require------------
(a) Food, moisture, proper pH
(b) Food, moisture, proper pH, proper temperature
(c) Food, proper temperature and time
(d) Food, moisture, proper pH, proper temperature and time

270. Insect and rodent infestation------------
(a) Is a serious health risk
(b) Is usually minor
(c) Does not always require professional examination
(d) Requires immediate evacuation of the food service facility

271. Too long a fermentation of sauerkraut may favour the growth of--------------
(a) *Lactobacillus lactis* (b) *Lactobacillus brevis*
(c) *Bacillus subtilis* (d) None of the above

272. Blackening in pickles occurs due to-------------
(a) The formation of hydrogen sulfide by bacteria
(b) Growth of Pencillium
(c) Both a and b
(d) None of the above

273. Vitamins, minerals and proteins in suitable amounts are given to body by---------

(a) Whole grain food (b) Fatty fish
(c) Plant oil (d) Nuts

274. According to food guide pyramid, fats oils and sweets should be used---------

(a) 2-3 servings (b) Frequently (c) Sparingly (d) 2-4 servings

275. Sources of proteins includes--------

(a) Fish (b) Poultry (c) Eggs (d) All of the above

276. A diet containing right amount of energy, carbohydrates, proteins, fats, fiber, vitamins, minerals and water to fulfill requirement of body is called-----------

(a) Nutrition (b) Balanced diet (c) Perfect diet (d) Food pyramid

277. Which of the following is NOT a typical dietary supplement form?

(a) Oil (b) Injection (c) Powder (d) Pills

278. Vitamin and mineral supplements would be highly appropriate for all types of individuals below, EXCEPT----------

(a) People who abuse alcohol or drugs
(b) People who eat restricted diets
(c) People who don't drink much milk or get sun exposure
(d) People who eat well and exercise often

279. HACCP is a system for-----------

(a) Controlling the food costs (b) Maintaining food safety
(c) Standardizing recipes (d) Controlling cooking quality

280. Fermentation is caused by -----------

(a) Bacteria (b) Viruses (c) Yeast (d) Parasites

281. Which of the following can be used to inhibit bacterial growth?

(a) Dehydration (b) High levels of salt (c) High levels of sugar (d) All of the above

282. Which of the following is a BAD food handling practice?

(a) Cleaning cutting boards and knives thoroughly after using with fresh meat
(b) Thoroughly cooking poultry before eating
(c) Prompt refrigeration of fresh and cooked foods in separate containers
(d) Storing cooked and fresh meat in the same container

283. What is a food borne illness caused by pathogens?

(a) Gastrointestinal (b) Poisonings (c) Infections (d) Parasites

284. Respiration helps in the conversion of food to------------

(a) Sugar (b) Nitrogen (c) Oxygen (d) Energy

285. Food spoilage is prevented by packing food in-----------

(a) Cellulose nitrate (b) Jute bags (c) Polythene bags (d) Nitro-packs

286. The demand function is the relationship between price and------------

(a) Net profit (b) Quantity purchased (c) Quantity sold (d) Gross income

287. The major component in milk is ------------

(a) Protein (b) Lactose (c) Fat (d) Water

288. The formation of free fatty acids causes a ----------- flavor in blue cheese.

(a) Sweet (b) Bitter (c) Salty (d) Rancid

289. Swelling of the can is caused primarily by------------

(a) Gas forming, anaerobic spore formers (b) Gas forming, aerobic spore formers

(c) Both a and b (d) None of the above

290. The possible health benefits of artificial sweeteners include-----------

(a) They help in weight loss and management

(b) They can be used in lieu of table sugar in case of diabetes

(c) They prevent dental cavities and tooth decay

(d) All of the above

291. A nutrient that can help lower blood pressure is------------

(a) Sodium (b) Fat (c) Protein (d) Potassium

292. The amount of nutrients a food contains in relationship to the number of calories it contains is-------------

(a) Caloric density (b) Nutrient density (c) Proportionality (d) Nutritive value

293. What drawbacks in saccharin forced the development of sweeteners with aspartame?

(a) Saccharin leaves a bitter after taste

(b) In the cooking process saccharin has to be added in the last cycle as it does not withstand heat

(c) It is not recommended for pregnant women as the placenta concentrates saccharin avidly

(d) All of the above

294. There are no Daily Values (DV) listed for trans fats and sugar because-----------

(a) They only appear in small quantities

(b) There is no need to control the intake of these items

(c) There is not enough information available to set reference values for them

(d) They are natural substances found in food

295. A claim on the label that describes a relationship between a food and a disease is called as-----

(a) Approved claim (b) Structure/function claim

(c) Nutrient claim (d) Health claim

296. Hypersensitivity reaction of the immune system to a component in a food is called ----------

(a) Food malabsorption (b) Food allergy

(c) Metabolic syndrome (d) Food intolerance

297. Symptoms of lactose intolerance include -------------

(a) Coughing (b) Difficult breathing (c) Skin rashes (d) Intestinal gas

298. At constant pressure, with increase of temperature, the dew point will-----------

(a) Increase (b) Remain unchanged (c) Decrease (d) None of the above

299. A positive blood text for food allergy must include ------------

(a) Antioxidants (b) Gluten (c) Enzymes (d) Antibodies

300. Quality as defined by the customer is ----------

(a) An unrealistic definition of quality

(b) A user-based definition of quality

(c) A manufacturing-based definition of quality

(d) A product-based definition of quality

301. The supply chain concept originated in what discipline?

(a) Marketing (b) Operations (c) Logistics (d) Production

302. Zero defects in manufacturing is------------

(a) A relevant goal only in electronic assembly

(b) Readily achievable in all areas

(c) The goal of TQM

(d) An unobtainable and misleading idea

303. All of the following costs are likely to decrease as a result of better quality except ----------

(a) Customer dissatisfaction costs (b) Maintenance costs

(c) Warranty and service costs (d) Scrap costs

304. Salting, as a preservative---------------

(a) Retards growth of *Staphylococcus aureus* (b) Plasmolyzes bacteria and fungi

(c) Is used to prevent growth of halophiles (d) All of the above

305. Concerning disorders due to vitamin deficiency or excess-----------

(a) Excess intake of vitamin A can cause severe toxic reactions

(b) Vitamin B_2 deficiency may cause severe muscular disorders

(c) Vitamin B_{12} deficiency results in microcytic anaemia

(d) Rickets can be prevented by large doses of vitamin C

306. Whole meal bread and cereals are rich in------------

(a) Protein, fibre, vitamin C (b) Carbohydrate, fibre, vitamin B

(c) Fat, iron, protein (d) Water, fatty acid, amino acids

307. Which nutrient is a concentrated form of energy?

(a) Fat (b) Carbohydrate (c) Protein (d) Vitamins

308. *Listeria monocytogenes* is a harmful microorganism found in---------

(a) Undercooked meat (b) Soft cheeses and unpasteurized milk

(c) Pork (d) Poultry

309. Which fact about Salmonella bacteria is true?
(a) Fruits, vegetables, and even processed foods may contain Salmonella
(b) Salmonella infection is more common in the summer
(c) Lightly cooked egg white and yolk can cause Salmonella infection
(d) All of the above

310. Nitrite prevents the growth of-------------
(a) *C. botulinum* (b) *C. perfringens* (c) *S. aureus* (d) Yeasts

311. The best way of cooking to maintain antioxidant activity in vegetables is by -----------
(a) Boiling (b) Steaming (c) Microwaving (d) All of the above

312. Which of the artificial sweeteners are approved by the FDA?
(a) Aspartame (b) Saccharin (c) Sucralose (d) All of the above

313. When would you say that the levels of cholesterol in the blood are high or the person has hypercholesterolemia?
(a) Cholesterol levels > 50 mg/dl (b) Cholesterol levels > 150 mg/dl
(c) Cholesterol levels > 200 mg/dl (d) Cholesterol levels > 240 mg/dl

314. Which of the following is NOT one of the types of information required by the FDA to appear on the labels of dietary supplements?
(a) A Supplement Facts panel listing serving size, number of tablets and related information
(b) A list of all the ingredients
(c) The contents of the bottle
(d) A list of all the endorsements

315. Which of the following is NOT a typical dietary supplement substance?
(a) Amino acid (b) Botanical (herb.
(c) Nutrient/micronutrient (d) Prescription medication

316. Creatine, caffeine, anabolic steroids, and growth hormones are all examples of what?
(a) Ergogenic aids (b) Health claims (c) Functional foods (d) Illicit substances

317. What is a functional food?
(a) One that provides high levels of nutrients above and beyond the daily recommendations
(b) One that is convenient and easily prepared
(c) One that protects against a number of potentially serious diseases or ailments
(d) One that has a positive effect on health beyond providing basic nutrients

318. In gel electrophoresis, why is the buffer important?
(a) It is the membrane that supports the matrix is the column
(b) It is the material that fractionates the chemicals in the sample
(c) It is the solvent in which the samples are dissolved
(d) It conducts the electric charge and controls the pH

319. Which of the following is not a typical supply chain member?
(a) Retailer's creditor (b) Wholesaler
(c) Reseller (d) Producer

320. TQM refers to -----------
(a) Total quantity management (b) Total quality management
(c) Total quality marketing (d) Total quotient management

321. Intermediaries play an important role in matching-----------
(a) Dealer with customer (b) Manufacturer to product
(c) Information and promotion (d) Supply and demand

322. Standard Operating Procedure (SOP) refers to-----------
(a) An optimal balance between possibilities realized and a framework of norms and values
(b) Doing the right thing right, right away, the first time
(c) Detailed, written instructions to achieve uniformity of the performance of a specific function
(d) A process of meeting the needs and expectations of the customers, both internal and external

323. Accreditation refers to one of the following------------
(a) Making assessment (b) Meeting predetermined standards
(c) Making care cost effective (d) Ensuring quality

324. Fermentation is the -----------
(a) Production of alcoholic beverages by microorganisms
(b) Mass, controlled culture of microbes to synthesize products
(c) Use of microbes in sewage and pollution control
(d) All of the above

325. The following are the antioxidants used in anti-aging products ----------
(a) Vitamins C and E (b) Coenzyme Q10
(c) Beta carotene (d) All of the above

326. Artificial sweeteners are found in------------
(a) Baked goods, puddings and candy (b) Soft drinks and powdered drink mixes
(c) Jams, jellies and dairy products (d) All of the above

327. Which type of lipoproteins protects the heart from coronary heart disease?
(a) Low Density Lipoproteins (LDL) (b) Very Low Density Lipoproteins (VLDL)
(c) Lp(a. Cholesterol (d) High Density Lipoproteins (HDL)

328. A synbiotic is a product that contains------------
(a) A probiotic with an antibiotic (b) A prebiotic with an antibiotic
(c) A prebiotic with a postbiotic (d) None of the above

329. The observation that no two substances produce the same fragmentation patterns under carefully controlled conditions is a unique feature of----------
(a) Mass spectrometry (b) Thin-layer chromatography
(c) Ultraviolet spectrophotometry (d) Electromagnetic radiation

330. The difference of wet bulb temperature and adiabatic saturation temperature of unsaturated mixture of any system is------------

(a) –ve (b) +ve (c) Zero (d) None of the above

331. Food heats up in a microwave oven primarily due to vibration of------------- molecules.

(a) Water (b) Fat (c) Protein (d) Carbohydrate

332. Using a process called -------------, liquid vegetable oils are changed to shortening and margarine.

(a) Hydrogenation (b) Oxidation (c) Saturation (d) Aeration

333. When fruits such as pears, apples, or bananas are cut or bruised, causes the cut surface to become discolored due to-------------

(a) Maillard reaction (b) Enzymatic browning

(c) Exposure to light (d) Catabolism

334. Chemical leavening agents such as baking soda and baking powder produce ------------ during baking to lighten or aerate baked goods.

(a) Carbon monoxide (b) Carbon dioxide

(c) Sodium bicarbonate (d) Steam

335. Which statement is true about flavouring materials spices, garlic etc. added to dill pickles?

(a) These markedly stimulate or inhibit the acid forming bacteria

(b) They may be source of considerable numbers of undesirable micro-organisms and cause off-fermentations or spoilage of pickles

(c) They may improve texture of the pickle

(d) None of the above

336. Shriveling in fermented pickles results from the physical effect of----------------

(a) Too strong salt solution (b) Too strong sugar solution

(c) Too strong vinegar solution (d) All of the above

337. Abnormal fermentation of cabbage may result in a cheese like odour caused by------------

(a) Propionic acid (b) Butyric acid

(c) Caprioc acid and Valeric acid (d) All of the above

338. During the production of sauerkraut, cabbage is ------------- to contribute to the aroma, flavor and color of sauerkraut.

(a) Fermented (b) Homogenized (c) Pasteurized (d) Lyophilized

339. Pressure canners used in the commercial manufacture of canned products are known as-------

(a) Steam blanchers (b) Retorts (c) Plate exchangers (d) Sublimators

340. Bioremediation uses -----------

(a) Microorganisms' natural capacities to break materials down

(b) Added oxygen, water, and nutrients to increase rates of degradation

(c) Organisms such as *Pseudomonas* and *Bacillus*

(d) All of the above

341. Water treatment typically requires three stages to make it potable from sewage. Which is the secondary stage?
(a) Skimming (b) Settling (c) Biodegradation (d) Filtration

342. Food fermentation include------------
(a) Leavening of bread (b) Pickling
(c) Alcohol production (d) All of the above

343. During malting, barley and other grains are broken down by ------------
(a) Heating to 95^{0}C (b) Lagering (c) Amylases (d) Yeasts

344. The "flash method" or "high temperature short time" method exposes fluids to-----------
(a) Heat below 100^{0}C (b) 62.3^{0}C for 30 minutes
(c) 134^{0}C for 1 second (d) 72^{0}C for 15 seconds

345. Chemical preservatives do NOT include -----------
(a) Organic acids (b) Sulfites (c) Starch (d) Alcohol

346. Combining the monosaccharide's glucose and fructose will yield-------------
(a) Cane sugar (b) Table sugar (c) Sucrose (d) All of the above

347. Complex carbohydrates include--------------
(a) Polysaccharides (b) Glycogen (c) Starches (d) All of the above

348. Nitrates maintain the red color of preserved meats and -------------
(a) Are among the most widely used preservatives
(b) Inhibit germination of botulism spores
(c) Maintain a high osmotic pressure to kill microorganisms
(d) Prevent mold

349. The production of substances in industrial microbiology occurs in the following sequence---
(a) Fermentation, downstream processing, removal of waste, inoculation
(b) Inoculation, downstream processing, fermentation, removal of waste
(c) Inoculation, fermentation, downstream processing, removal of waste
(d) Removal of waste, inoculation, fermentation, downstream processing

350. Food poisoning may be caused by -------------
(a) Heavy metals (b) Chemicals (c) Drugs (d) All of the above

351. Sodium benzoate is used in soft drinks primarily to inhibit--------------
(a) Rancidity (b) Color deterioration (c) Mold growth (d) Flavor breakdown

352. There are -------------- principles of HACCP.
(a) 3 (b) 5 (c) 7 (d) 9

353. Energy lost when water molecules form ice crystals is called -------------
(a) Specific heat (b) Latent heat (c) Heat of fusion (d) Heat of vaporization

354. A retort is a piece of equipment used for-------------
(a) Frying (b) Drying (c) Canning (d) Baking

355. The purpose of using a leavening agent such as baking soda or baking powder in cakes and cookies is to provide a source of--------------
(a) Sodium dioxide (b) Carbon monoxide (c) Sodium monoxide (d) Carbon dioxide

356. Fruits and vegetables are primarily composed of------------
(a) Carbohydrates (b) Water (c) Protein (d) Fiber

357. Regarding dietary requirements of the major foodstuffs-----------
(a) Adults can synthesize all of the amino acids found in proteins
(b) Children who have a protein deficient diet may develop kwashiorkor
(c) Oedema is a characteristic feature of marasmus
(d) People whose diets are low in fats often suffer from fatty acid deficiency diseases

358. Sugar is a natural sweetener. What is the sweetening agent in it?
(a) Basil (b) Zerovia (c) Stevia (d) Parsley

359. What is the process of converting liquid fat into solid and semisolid fats known as?
(a) Saponification (b) Emulsification (c) Hydrogenation (d) Deodorization

360. A food that has a high glycemic index is-------------
(a) Honey (b) Pasta (c) Margarine (d) All of the above

361. Vegetables and fruit provide----------
(a) Fibre, fat and protein (b) Vitamin C and A, fibre and many minerals
(c) Vitamin D (d) Creatine

362. The formula for Body Mass Index is-----------
(a) Height divided by weight (b) Weight divided by height
(c) Weight divided by height squared (d) Height divided by weight squared

363. Which food type in our diet provides us with the maximum amount of energy?
(a) Fats or Lipids (b) Carbohydrates (c) Proteins (d) Vitamins and Minerals

364. Protein is made up of which molecule?
(a) Lipid (b) Carbon and oxygen (c) Amino acid (d) Oxygen and hydrogen

365. All the drinks listed below have antioxidant properties but which is considered the most potent of the four?
(a) Apple juice (b) Orange juice (c) Green tea (d) Red wine

366. The immobilized technique involving chemical method is----------------
(a) Covalent bond formation dependent (b) Non-covalent bond formation dependent
(c) Both a and b (d) Ionic bond formation dependent

367. Which of the following statements about amines is incorrect?
(a) They react with acids to form salts
(b) Aliphatic amines are more basic than aromatic amines
(c) Primary amines are more basic than tertiary amines
(d) The biological oxidation of amines is usually carried out by monoamine oxidases

368. Probiotics are effective in which of the following diarrheas-------------
    (a) Rotavirus diarrhea
    (b) Antibiotic-associated diarrhea
    (c) Radiation-induced diarrhea
    (d) All of the above
369. In reversed phase HPLC, there is a------------
    (a) Non polar solvent/polar column
    (b) Polar solvent/non-polar column
    (c) Non polar solvent/non-polar column
    (d) Any of the above
370. Increasing the acid content of a food is effective in preventing the growth of-----------
    (a) *Clostridium botulinum*
    (b) Salmonella
    (c) Molds
    (d) Yeasts
371. In aseptic packaging, what is sterilized?
    (a) The food product
    (b) The foodservice worker's hands
    (c) The packaging
    (d) Both a and c
372. The food-borne illness that can be caused by a foodservice worker coughing or sneezing on food is-------------
    (a) *Staphylococcus aureus*
    (b) *Clostridium perfringens*
    (c) *Salmonella*
    (d) *Trichanella*
373. Which of the following is NOT a good food safety practice?
    (a) Cooking chicken to an internal temperature of 180^0F
    (b) Cooling leftovers rapidly in shallow pans
    (c) Avoiding sneezing and coughing over foods
    (d) Serving meals with raw fish
374. One of the most common food-borne illnesses from undercooked poultry and eggs is--------
    (a) *Trichanella*
    (b) *Salmonella*
    (c) *Clostridium perfringens*
    (d) *Clostridium botulinum*
375. A virus found in shellfish from contaminated water and foods that are handled by infected individuals then eaten raw is-------------
    (a) Influenza
    (b) Hepatitis C
    (c) Hepatitis A
    (d) Rhinovirus
376. Safe methods for thawing foods include all of the following EXCEPT-----------
    (a) Thawing under cold running water
    (b) Thawing in the refrigerator
    (c) Thawing in the microwave
    (d) Thawing in a sink of warm water
377. Which governmental agency is responsible for enforcing standards for wholesomeness and quality of grains, produce, meat, poultry, milk, and eggs produced in the US, through inspection and grading------------
    (a) USDA
    (b) FDA
    (c) CDC
    (d) EPA
378. Chromatography can be used to----------
    (a) Form mixtures
    (b) Change mixture compositions
    (c) Separate mixtures into pure substances
    (d) All of the above

379. Which of the following are essential for body growth and formation of new cells?
(a) Carbohydrates (b) Protein (c) Minerals (d) Vitamins

380. The component of food which is the major energy supplier is-------------
(a) Cellulose (b) Protein (c) Carbohydrate (d) Amino acid

381. Which of the following are the rich sources of protein?
(a) Iron and calcium (b) Meat, fish, eggs (c) Rice and wheat (d) Fruits and vegetables

382. The most common cause of food-borne illness is the presence of --------------
(a) Bacteria (b) Chemical preservatives
(c) Viruses (d) Molds

383. Ancient methods of food preservation include------------
(a) Pasteurizing and sterilizing (b) Canning, blanching, and irradiating
(c) Freezing and boiling (d) Drying, smoking, and fermenting

384. A biological method of food preservation is------------
(a) Drying (b) Freezing (c) Fermentation (d) Adding salt

385. The organisms used in the process of fermentation----------
(a) Metabolize all the oxygen in food
(b) Produce water in the food
(c) Utilize all the nutrients in a food
(d) Produce products, such as acids, that inhibit the growth of other organisms

386. A danger of using home-canned food if the processing has been inadequate is-----------
(a) Cancer (b) Botulism
(c) Excessive nutrient loss (d) Excessive tin in the juice

387. Keeping food above 140o F (60°C.-------------
(a) Increases microbial growth (b) Slows chemical deterioration
(c) Stops chemical reactions (d) Reduces the growth of pathogens

388. Food additives may be used to-------------
(a) Destroy nutrients (b) Disguise faulty products
(c) Deceive customers (d) Enhance appearance

389. Antioxidants that prevent rancidity include-------------
(a) Smoke (b) Salt (c) BHT and BHA (d) Sugar

390. A combination of paper chromatography and electrophoresis involves------------
(a) Partition chromatography (b) Electrical mobility of the ionic species
(c) Both a and b (d) None of the above

391. The factor that is least important to control in order to limit food-borne illness is----------
(a) Presence of pesticides (b) Presence of microbes
(c) Temperature of food (d) Time of incubation

392. Substances used to preserve foods by lowering the pH are------------
    (a) Salt and sugar (b) Smoke and irradiation
    (c) Baking powder and soda (d) Vinegar and citric acid
393. What are the common antioxidant substances?
    (a) Beta carotene (b) Lutein (c) Lycopene (d) All of the above
394. Which of these is a common food allergen?
    (a) Soya (b) Peanuts, hazelnuts, almonds, walnuts
    (c) Seafood (d) All of the above
395. Which of the following food can be used as a "gluten free diet" to prevent "gluten allergy"?
    (a) Wheat (b) Rice (c) Barley (d) Rye
396. Which of the following is not a natural sugar?
    (a) Glucose (b) Fructose (c) Saccharin (d) Xylitol
397. Which parameter is used to calculate body mass index of an individual?
    (a) Weight (b) Height
    (c) Abdominal circumference (d) Both height and weight
398. HDL-cholesterol-----------
    (a) Protects against atherosclerosis (b) Worsens atherosclerosis
    (c) Has no effect on atherosclerosis (d) None of the above
399. Most important method of preventing food borne illness is -----------
    (a) Cooking every food for at least 15-20 min.
    (b) Sanitizing the kitchen every time
    (c) Washing hands thoroughly and frequently
    (d) Using paper plates and cups for eating
400. Which statement about food poisoning bacteria is not true?
    (a) The food contaminated with bacteria looks, tastes and smells perfectly normal
    (b) Freezing food kills the bacteria
    (c) Damp cotton towels are a breeding ground for food poisoning bacteria
    (d) Food poisoning bacteria thrive best at temperatures between 5-63°C (40-140°F)
401. What are the four bases in DNA?
    (a) First, second, third and fourth
    (b) Adenine, guanine, cytosine, uracil
    (c) Nitrogen, hydrogen, carbon and phosphate
    (d) Adenine, guanine, cytosine, thymine
402. Heat exchanger tubes are never made of------------
    (a) Plain carbon steel (b) Lead
    (c) Stainless steel (d) Copper

403. ------------ is the mobile phase leaving the column.
(a) Eluate (b) Eluent (c) Analyte (d) Both a and b

404. Which of the following is the "energy currency" molecule that our bodies create?
(a) DNA (b) ATP (c) Glucose (d) Lipids

405. To increase fiber intake which of the following foods should one consume?
(a) White bread (b) Peas (c) Dinner rolls (d) None of the above

406. The requirement for -----------increases the cost of supercritical fluid extraction compared to conventional liquid extraction.
(a) High pressures (b) High vacuum (c) High temperature (d) All of the above

407. Which of the following foods is a reliable food source of vitamin E?
(a) Milk (b) Almonds (c) Spinach (d) Pumpkin

408. Iodine value measures--------------
(a) Degree of unsaturation (b) Amount of carbon present
(c) Degree of saturation (d) No. of iodine present

409. Food additives widely used without apparent ill effects are in the ---------------list.
(a) FDA (b) GRAS (c) Additive safety (d) Delaney

410. Obesity is a risk factor for which condition?
(a) Heart diseases (b) Arthritis
(c) Sleep apnea syndrome (d) All of the above

411. Which of the foodstuff can lower blood cholesterol level?
(a) Banana (b) Spinach (c) Garlic (d) All of the above

412. Which of the cooking medium is a rich source of PUFA?
(a) Palm oil (b) Coconut oil (c) Groundnut oil (d) Safflower oil

413. Which of the following has the best chemical score?
(a) Wheat (b) Egg (c) Milk (d) Pulses

414. Which of the following has the highest glycemic index?
(a) Ice cream (b) Cucumber (c) Bread (d) Dextrose

415. All except one are true about the significance of dietary fiber?
(a) It promotes peristalsis (b) Reduces cholesterol absorption
(c) Increases glycemic index (d) Acts as an antioxidant

416. Which of the following is not a component of dietary fiber?
(a) Cellulose (b) Pectin (c) Lignin (d) Agar

417. Which of the following statements about alcohols is incorrect?
(a) Tertiary alcohols have lower boiling points than primary alcohols with an equivalent molecular weight
(b) Alcohols undergo nucleophilic substitution

(c) Tertiary alcohols undergo dehydration more readily than primary alcohols

(d) Tertiary alcohols are metabolized in the body to ketones

418. A trickle filter system is associated with which of the following steps of wastewater treatment?

(a) Chemical treatment (b) Sludge treatment

(c) Primary treatment (d) Secondary treatment

419. All of the following are aspects of continuous flow production in industrial fermentation EXCEPT---------------

(a) The microbe secretes its product into the medium

(b) Wastes are constantly removed

(c) The fermentation product is harvested all at once

(d) Fresh nutrients are constantly added

420 The government agency responsible for ensuring that meat and poultry are safe and wholesome for consumption is the------------

(a) Food and Drug Administration

(b) United States Department of Agriculture

(c) Department of Health and Human Services

(d) Animal and Plant Health Inspection Service

421. --------------reacts with amino acids when milk is heated to contribute to the tan color and slightly caramelized flavor of cooked milk products.

(a) Lactose (b) Casein (c) Whey (d) Milk fat

422. An additive that can keep a compound, mixture or solution from changing its form or chemical nature is called a----------------

(a) Antioxidant (b) Buffer (c) Stabilizer (d) Preservative

423. Flavor is sensed by taste buds which are sensory organs located on parts of the tongue. The taste buds on the sides of the tongue respond to -------------flavors.

(a) Sweet (b) Bitter (c) Salty (d) Sour

424. Which of the following is a primary aromatic amine?

(a) Aniline (b) Anisole (c) Anthracene (d) Adrenaline

425. Which of the following statements about quaternary ammonium compounds is incorrect?

(a) They are neutral crystalline solids (b) They are insoluble in water

(c) They are insoluble in lipids (d) They are ionic molecules

426. Soft pickles are made so by pectolytic enzymes mostly from----------------

(a) Moulds (b) Yeasts (c) Bacteria (d) None of the above

427. Hollow pickles are obtained as a result of----------------

(a) Loose packing in the vat (b) Too strong or too weak brine

(c) Both a and b (d) None of the above

428. What are probiotics?
(a) Killed microorganisms that provide health benefits
(b) Live microorganisms that provide health benefits
(c) Bacterial toxins that provide health benefits
(d) Live microorganisms that could be harmful

429. The key areas to a balanced approach to weight loss includes ------------
(a) Diet and exercise
(b) Diet, good sleep and happiness
(c) Diet, exercise and motivation
(d) Exercise and good sleep

430. Basal metabolic rate (BMR) is ----------
(a) The type of metabolism you are born with
(b) The minimum amount of calories required to sustain life when at rest
(c) The length of time required to lose a certain amount of fat
(d) The rate at which the body produces fat

431. Avocado is a good inclusion in the weight loss diet because this fruit is high in----------
(a) Calcium (b) Healthy fats (c) Iron (d) Vitamin C

432. The technique first described to determine the incipient spoilage in meat was------------
(a) Homogenate Extract Volume (HEV) (b) Agar Plate Count (APC.
(c) Extract Release Volume (ERV) (d) None of the above

433. Molds causing spoilage of eggs include species of-----------
(a) *Cladosporium* (b) *Mucor* (c) *Thamnidium* (d) All of the above

434. Vacuum packaged meats are spoiled by----------
(a) *B. thermosphacta* (b) *Lactobacilli* (c) Both a and b (d) None of the above

435. Which of the following foods is produced by fermentation involving lactic acid bacteria?
(a) Beer (b) Coffee (c) Yoghurt (d) Vinegar

436. Paper chromatography separates molecules according to their-----------
(a) Molecular size (b) Polarity (c) Solubility (d) Matrix

437. What is the primary reason for blanching foods?
(a) Cleans the food (b) Inactivates enzymes found in food
(c) Prevents pest infestation (d) Prevents food from drying out

438. What is the operating principle behind oven drying for determining moisture content of foods?
(a) Colour change is measured (b) Loss of weight represents loss of water
(c) Change in refractive index is measured (d) Change in light absorbance is measured

439. Which gas causes fruits to ripen?
(a) Carbon monoxide (b) Nitrogen
(c) Propane (d) Ethylene

440. Which of the following chemicals is a solvent used for testing fat content?
(a) Hydrochloric acid (b) Sodium hydroxide
(c) Amylase (d) Ether

441. Which microorganism is commonly associated with fecal contamination?
(a) *Clostridium botulinum* (b) *Campylobacter jejuni*
(c) *Bacillus cereus* (d) *Trichinella spiralis*

442. Which industrial processing method is most effective for making dried potato flakes?
(a) Drum drying (b) Sun drying (c) Spray drying (d) Osmotic dehydration

443. Which of the following analytical methods can be used to distinguish flavour compounds?
(a) Hydrometry (b) Near infrared spectroscopy
(c) Polarimetry (d) Gas chromatography

444. Which of the following microorganisms cannot tolerate oxygen?
(a) *Clostridium botulinum* (b) *Staphylococcus aureus*
(c) *Penicillium roquefortii* (d) *E. coli*

445. Protein derived from animal sources is typically--------------
(a) Incomplete until cooked (b) Complete
(c) Incomplete (d) Nearly complete

446. What happens to the boiling point of water when it is heated at high altitudes?
(a) Increases (b) Decreases
(c) Stays the same (d) Water doesn't boil at high altitudes

447. Canadian and American laws prescribe how ingredients should be listed on labels. What is the general stipulation with respect to the order that ingredients are listed?
(a) By alphabetical order
(b) By ascending order of proportion by weight
(c) By descending order of proportion by weight
(d) By descending order of proportion by volume

448. Which of the following methods is a quick test for sugar content during the early stages of the brewing process for beer?
(a) Hydrometry (b) Babcock test (b) Babcock test (d) Soxhlet extraction

449. Which of the following processing methods involves heating foods at high temperatures for short periods of time in order to reduce the risk of food poisoning?
(a) Blanching (d) Soxhlet extraction (c) Irradiation (d) Pasteurization

450. What causes sliced apples to turn brown?
(a) Caramelization (b) Staling (c) Enzyme activity (d) Protein degradation

451. Which of the following does not have antimicrobial activity?
(a) Chlorophyll (b) Organic acids (c) Spice extracts (d) Hydrogen peroxide

452. From which plant source is gluten derived?

(a) Soybean (b) Corn (c) Rice (d) Wheat

453. Fermentation is the-------------

(a) Production of alcoholic beverages by microorganisms

(b) Mass, controlled culture of microbes to synthesize products

(c) Use of microbes in sewage and pollution control

(d) All of the above

454. ------------ is the separation technique used for desalination of sea water.

(a) Reverse osmosis (b) Thermal diffusion (c) Absorption (d) Adsorption

455. Water treatment typically requires three stages to make it potable from sewage. Which is the secondary stage?

(a) Skimming (b) Settling (c) Biodegradation (d) Filtration

456. Industrial microbiologists use "tricks" to increase the amount of chosen end product; this may include---------------

(a) Manipulating the growth environment

(b) Using small scale containers to concentrate production

(c) Selecting microbial strains that lack a feedback system

(d) Both a and c

457. If the voltage of the power supply in gel electrophoresis is reduced, how will this affect the results?

(a) The pH will drop and destroy the samples

(b) The molecules will move slower

(c) The molecules will move faster

(d) The voltage of the power supply has no real effect on gel electrophoresis

458. Why food irradiation is nicknamed "cold pasteurization"?

(a) Because it destroys pathogens without cooking the food

(b) Because it was originally invented by Louis Pasteur

(c) Because irradiated food must be kept frozen

(d) All of the above

459. Chromatography is especially useful for-------------

(a) Foods (b) Drugs (c) Both a and b (d) None of the above

460. Relative humidity is the ratio of the--------------

(a) Partial pressure of the vapour to the vapour pressure of the liquid at gas temperature

(b) Actual humidity to saturation humidity

(c) Partial pressure of the vapour to the vapour pressure of the liquid at room temperature

(d) None of the above

461. Drying operation under vacuum is carried out to------------
(a) Dry those materials which have very high unbound moisture content
(b) Reduce drying temperature
(c) Increase drying temperature
(d) Dry materials having high bound moisture content

462. Which one of the following statements is correct?
(a) Starch solution is coloured blue/black
(b) Benedict's solution is colourless
(c) All sugars are reducing sugars
(d) Benedict's solution changes colour to red in the presence of a reducing sugar

463. Which one of the following methods is used to separate the colours in food dyes?
(a) Sieving (b) Decanting (c) Chromatography (d) Evaporation

464. Mass transfer rate between two fluid phases does not necessarily depend on the ----------- of the two phases.
(a) Physical properties (b) Chemical properties
(c) Interfacial area (d) Degree of turbulence

465. Heat in BTU necessary to increase the temperature of 1 lb of gas and its accompanying vapour by 1°F is called the-----------
(a) Latent heat (b) Specific heat (c) Humid heat (d) Sensible heat

466. An ion exchange resin is composed of------------
(a) Polymeric network (b) Ionic functional groups
(c) Counter ions (d) All of the above

467. Which of the operation does not come under upstream processing?
(a) Media preparation (b) Inoculum development
(c) Effluent treatment (d) Storage of raw material

468. Chromatography is based on the------------
(a) Different rate of movement of the solute in the column
(b) Separation of one solute from other constituents by being captured on the adsorbent
(c) Different rate of movement of the solvent in the column
(d) Any of the above

469. Which one of the following is NOT a function of water in the human body?
(a) Helps eliminate wastes (b) It regulates body temperature
(c) Helps in the formation of bones (d) It transports nutrients around the body

470. A musty or muddy odor of the fish is attributed to-------------
(a) The growth of *Streptomyces* species in the mud at the bottom of the body of water
(b) The mud at the bottom of the body of water
(c) The growth of *Pseudomonas* species in the mud at the bottom of the body of water
(d) None of the above

471. In chilled shrimp ---------------- is chiefly responsible for spoilage.
(a) *Achromobacter* (b) *Pseudomonas*
(c) *Micrococcus or Bacillus species* (d) *Molds or yeasts*

472. *Saccharomyces cerevisiae* is used in the production of which of the following?
(a) Cheddar cheese (b) Yogurt (c) Leavened breads (d) Pickles

473. What's the healthiest part of an egg?
(a) Yolk (b) White (c) Yolk and white (d) None of the above

474. Which fish has the healthiest fats?
(a) Cod (b) Flounder (c) King mackerel (d) Salmon

475. Golden rice is a genetically modified crop plant where the incorporated genes are meant for biosynthesis of ----------
(a) Vitamin A (b) Vitamin C (c) Vitamin B (d) Beta-carotene

476. In Fourier's law, the proportionality constant is called the-----------
(a) Thermal conductivity (b) Thermal diffusivity
(c) Heat transfer co-efficient (d) Stefan-Boltzman constant

477. In column chromatography, why is the buffer important?
(a) It is the membrane that supports the matrix is the column
(b) It is the material that fractionates the chemicals in the sample
(c) It controls the pH of the sample so the shapes of the molecules won't change
(d) It conducts the electric charge

478. Which of the following statements is incorrect regarding transport proteins?
(a) They are present in cell membranes
(b) They serve to carry polar molecules across the hydrophobic cell membrane
(c) They are required to transport amino acids across cell membranes
(d) They are required to transport hydrophobic steroids across cell membranes

479. Which nuts have healthy fats, fiber, and protein?
(a) Almonds (b) Peanuts (c) Walnuts (d) All of the above

480. What can oatmeal help to control?
(a) Cholesterol levels (b) Blood sugar levels
(c) Body weight (d) All of the above

481. Which one of these foods is likely to contain the most bacteria?
(a) Cooked chicken (b) Tinned cream (c) Frozen raw chicken (d) Bottled mayonnaise

482. High performance liquid chromatography (HPLC. cannot be used to ----------
(a) Separate types of organic pesticides
(b) Determine the mercury content of a fish sample
(c) Identify the various pigments from a leaf extract
(d) Determine the caffeine content of coffee samples

483. At work, the best way to dry hands after washing is-----------

(a) Using a warm air dryer (b) Give them a shake

(c) Using a cotton towel (d) Using a paper towel

484. To determine the amount of free water available for microbes to use in a food system, a food scientist would measure the ------------ of that foo(d)

(a) Water activity (b) Percent water (c) pH (d) Brix

485. When peanuts are ground to make peanut butter, a ------------- is adde(d) This keeps the peanut oil from separating out to the top of the jar during storage.

(a) Caking agent (b) Stabilizer (c) Humectant (d) Antioxidant

486. A food that would be rich in omega-3 fatty acids would be-------------

(a) Fatty fish (b) Lard (c) Olive oil (d) Butter

487. Cheese curd is primarily composed of coagulated ---------------

(a) Protein (b) Fat (c) Carbohydrate (d) Lactose

488. Sodium benzoate is used as a preservative in soft drinks to inhibit growth of------------

(a) Bacteria (b) Molds (c) Yeasts (d) Viruses

489. When proteins begin to break down in meat, the process is called------------

(a) Proteolysis (b) Lipolysis (c) Glycolysis (d) Hydrolysis

490. Which list of vegetables is considered "free foods" because they don't have a significant amount of carbohydrates in them------------

(a) Sweet potatoes, corn, broccoli (b) Mushrooms, bell peppers, cucumbers

(c) Peas, kidney beans, potatoes (d) All of the above

491. According to federal health officials and heart experts, what percentage of our daily calories should come from carbohydrates?

(a) Less than 15% (b) Less than 35%

(c) Between 45 and 65% (d) Between 70 and 80%

492. What does the glycemic index measure?

(a) The ratio of carbohydrates to fat in a particular food

(b) How much and how quickly sugar gets into your bloodstream after you eat food that contains carbohydrates

(c) Weight (in kilograms) divided by height (in meters)

(d) The carbohydrate-to-sugar ratio of a food item

493. The "building blocks" of protein are called----------

(a) Essential fatty acids (b) Amino acids

(c) Fibrous tissues (d) Complementary proteins

494. Which of the following is NOT an intrinsic factor in food spoilage?

(a) pH (b) Moisture content (c) Physical structure (d) Temperature

495. In reverse phase chromatography, the stationary phase is made--------------

(a) Non-polar (b) Polar

(c) Either non-polar or polar (d) None of the above

496. What are two primary problems associated with protein deficiency in the diet?

(a) Irregular heartbeat and intestinal problems

(b) Allergies and sleep problems

(c) Anemia and weight fluctuations

(d) Loss of bone mass and protein-energy malnutrition

497. Which of the following terms describes organisms that thrive in the cold?

(a) Mesophiles (b) Thermophiles (c) Psychrophiles (d) Aerophiles

498. The basis of the technique of chromatography for separating components of a mixture is--------

(a) The absorption of infrared radiation by the components

(b) The interaction of the components with both stationary and mobile phases

(c) The deflection of charged particles in a magnetic field

(d) The differing movement of particles of different mass in an electric field

499. In a native PAGE, proteins are separated on the basis of--------------

(a) Net negative charge (b) Net charge and size

(c) Net positive charge size (d) Net positive charge

500. The hormone responsible for setting basal metabolic rate is---------

(a) Melanin (b) Melatonin (c) Thyroxine (d) Glucagon

501. The strongest form of intermolecular bonding that could be formed involving the residue of the amino acid serine is--------------

(a) Ionic bond (b) Hydrogen bond

(c) Van der Waals interactions (d) None of the above

502. Despite efforts to eliminate spoilage organisms during canning, sometimes canned foods are spoiled. This may be due to ----------

(a) Spoilage before canning

(b) Under processing during canning

(c) Leakage of contaminated water through can seams during cooling

(d) All of the above

503. Sodium nitrite is responsible for -------------

(a) Protecting against botulism (b) Reducing rate of spoilage

(c) Maintenance of red color in meat (d) All of the above

504. Which one of the following compounds does NOT cause hardness in water?

(a) Calcium chloride (b) Sodium chloride (c) Magnesium sulfate (d) Magnesium chloride

505. All of the following are associated with beer production EXCEPT---------------

(a) Sulfur dioxide (b) Hops (c) Wort (d) Malt

506. Which of the following assays could not be performed by gas chromatography?
   (a) Characterization of volatile oils
   (b) Measurement of drugs and metabolites in biological fluids
   (c) Characterization of raw materials for drug synthesis
   (d) Analysis of intravenous sodium chloride infusion
507. Chromatography refers to the------------
   (a) Concentration of chromium in a substance
   (b) Mass of chromium
   (c) Testing of additives
   (d) Testing of the alkalinity or acidity
508. Gel electrophoresis separates molecules according to their------------
   (a) Wavelengths of light absorbed (b) Polarity
   (c) Solubility (d) Matrix
509. Lack of enough potassium results in which problem?
   (a) Rickets (b) Tendency to hemorrhage
   (c) Osteomalacia (d) Muscle weakness
510. Which of the following terms refers to the arrangement of different protein subunits in a multi protein complex?
   (a) Primary structure (b) Secondary structure
   (c) Tertiary structure (d) Quaternary structure
511. The most important bacterium developing acidity in both low and high salt brines is-------------
   (a) *L. lactis* (b) *L. plantarum* (c) *L. brevis* (d) *B. subtilis*
512. The elimination of the scum of yeasts in the brine during the fermentation include-------------
   (a) Agitation of the surface (b) Addition of the mustard oil
   (c) Addition of the sorbic acid (d) All of the above
513. Which is NOT true of the bacterium produced by *Streptococcus lactis*?
   (a) A small hydrophobic protein
   (b) Nontoxic to humans
   (c) Affects mainly gram-negative bacteria
   (d) Improves inactivation of *C. botulinum* during the canning process
514. Absorption accompanied by heat evolution results in------------
   (a) Increased capacity of the absorber (b) Decrease in equilibrium solubility
   (c) Increase in equilibrium solubility (d) None of the above
515. Fractional solvent extraction------------
   (a) Results in low interfacial tension (b) Employs two solvents
   (c) Employs only one solvent (d) None of the above

516. The main reservoir of *Staphylococcus aureus* is-----------
(a) Human feces
(b) Human nasal cavity
(c) Human red blood cells
(d) All of the above

517. Which type of fermentation is used to produce yoghurt?
(a) Mesophilic
(b) Thermophilic
(c) Therapeutic
(d) Yeast-lactic fermentations

518. A compound that has little or no flavor itself but is added to food to assist or boost the primary flavor of the food to which it is added is a-----------
(a) Processing aid (b) Humectants (c) Stabilizer (d) Flavor enhancer

519. When a food processing plant is cleaned at the end of a production day, the order of cleanup is-
(a) Rinse, clean with detergent, dry pick up, rinse, sanitize
(b) Clean with detergent, rinse, sanitize, rinse, dry pick up
(c) Dry pick up, rinse, clean with detergent, rinse, sanitize
(d) Dry pick up, rinse, clean with detergent, sanitize, rinse

520. When water is used as an ingredient in food formulations, it must be -----------
(a) Soft water (b) Potable water (c) Hard water (d) Purified water

521. -------------is an ingredient used in food to slow the reaction of lipids forming free radicals leading to oxidative rancidity in food.
(a) Butylated hydroxyanisole
(b) Sodium caseinate
(c) Potassium sorbate
(d) Disodium inosinate

522. The brownish color of aerobically packaged ground beef that has been stored in a refrigerator for several days is due to -------------
(a) Deoxymyoglobin (b) Metmyoglobin (c) Myoglobin (d) Oxymyoglobin

523. Which of the following refers to the addition of microorganisms to the diet in order to provide health benefits beyond basic nutritive value?
(a) Antibiotics (b) Adjuvant (c) Prebiotics (d) Probiotics

524. Food handlers must always wash their hands ---------
(a) Before starting work
(b) Switching between handling raw and ready-to-eat food
(c) After going to restrooms
(d) All of the above

525. The presence or growth of microorganism is a type of-----------
(a) Physical hazard
(b) Chemical hazard
(c) Biological hazard
(d) Acidic hazard

526. Potentially hazardous hot foods must be maintained at an internal temperature of --------- or higher to prevent rapid bacterial growth.
(a) 145°F (b) 135°F (c) 160°F (d) 180°F

527. The only correct way to ensure the proper strength of a chemical sanitizer is to------------
(a) Measure it using correct test strips
(b) Change the solution every 8 hours
(c) Change the solution at the end of each work shift
(d) None of the above

528. Bacteria grow best in the temperature zone known as the "danger zone," which includes temperatures between-----------
(a) 0°F and 100°F (b) 32°F and 220°F (c) 41°F and 135°F (d) 41°F and 220°F

529. Which of the following is not an approved method to thaw potentially hazardous foods?
(a) In a microwave oven (b) During cooking process
(c) Under cool running water (d) On a clean counter, at the room temperature

530. Which of the following is approved for food-contact surfaces?
(a) Lead (b) Copper (c) Stainless steel (d) Galvanized metal

531. Foods that are decomposed or produced under unsanitary conditions are-----------
(a) Mislabeled (b) Potable
(c) Can be consumed within expiration date (d) Adulterated

532. HPLC stands for---------------
(a) High Pressure Liquid Chromatography (b) High Performance Liquid Chromatography
(c) Both a and b (d) Highly Placed Liquid Chromatography

533. In an SDS-PAGE------------
(a) Proteins are denatured by the SDS
(b) Proteins have the same charge-to-mass ratio
(c) Smaller proteins migrate more rapidly through the gel
(d) All of the above

534. Milk, butter and cheese contain cholesterol because they are-----------
(a) Homogenized (b) Animal products (c) Processed (d) High in fat

535. Most vitamins--------------
(a) Function as coenzymes in the body (b) Are carbohydrates
(c) Have only one function in the body (d) Are used as building blocks for the body

536. The process in which glucose is converted to a storage form----------
(a) Results in beta oxidation (b) Is called glycogenolysis
(c) Occurs in the liver (d) Requires vitamin A

537. Which of the following is the amount of energy the body needs to maintain life?
(a) MR (b) TMR (c) BMI (d) BMR

538. Which of the following terms refers to the order in which amino acids are linked together in a protein?
(a) Primary structure (b) Secondary structure
(c) Tertiary structure (d) Quaternary structure

539. Which of the following was the first amino acid to be produced commercially?
(a) L-glutamic acid (b) L-lysine (c) L-cystine (d) L-methionine

540. Pectinase, an enzyme used for increasing yield and for clarifying juice is obtained by------------
(a) *A. oryzae* (b) *A. niger* (c) *A. flavus* (d) All of the above

541. In anaerobic environment, the sugar in dough is converted into--------------
(a) Glucose (b) Water (c) Alcohol (d) Carbon monoxide

542. Separation of a mixture of two gases by absorption in the liquid solvent depends upon the difference in their-----------
(a) Relative volatility(b) Viscosity (c) Density (d) Solubility

543. Molality is defined as the number of gm moles of solute per ---------- of solvent.
(a) kg (b) litre (c) ml (d) gm

544. Which of the following is not a batch drier?
(a) Agitated pan drier (b) Truck drier
(c) Vacuum shelf drier (d) Fluidized bed drier

545. Which of the following is known as 'Flavr Savr'?
(a) Specific variety of pesticide (b) Breed of chicken
(c) Transgenic tomato (d) Toxic insecticidal protein

546. Which of the following statements is false regarding food safety?
(a) Failure to correct critical violations can result in possible closure of the food establishments
(b) Food establishments that have a manager or employee who is certified in food safety receive no inspections from local health departments
(c) Food safety managers should conduct periodic self-inspections to assure food safety practices
(d) A routine inspection can be done any time during regular or reasonable business hours

547. The formation of glucose from substrates other than carbohydrate is termed--------------
(a) Glycolysis (b) Gluconeogenesis (c) Lipolysis (d) Proteolysis

548. Low blood sugar is termed as--------------
(a) Hyperglycemia (b) Hypoglycemia (c) Hyperthermia (d) Hypothermia

549. What temperature is food frozen at?
(a) - 4^0C (b) - 18^0C (c) - 25^0C (d) - 40^0C

550. The heat transfer co-efficient in film type condensation is ----------that for drop wise condensation.
(a) Half (b) Greater than (c) Lower than (d) Is same as

551. Most important species of vinegar bacteria causing sliminess is---------------
(a) *Acetobacter aceti* (b) *Saccharomyces carlsbergiensis*
(c) *Lactobacillus lactis* (d) None of the above

552. Which of the following is not an ale type of beer?
(a) Weiss beer (b) Porterr (c) Stout (d) Pilsener

553. A vegetarian diet that includes fish, dairy products, and eggs, along with plant foods, is known as --------------- diet.
(a) Ovolactovegetarian (b) Lactovegetarian
(c) Pesco-vegetarian (d) Vegan vegetarian

554. Which of the following decreases during evaporative cooling process with recirculated water supply?
(a) Relative humidity (b) Wet bulb temperature
(c) Partial pressure of vapour (d) None of the above

555. Which of the following conditions is needed for bacteria to grow?
(a) Vinegar (b) Salt (c) Water (d) All of the above

556. What method of preservation uses rays to destroy enzymes in food?
(a) Canning (b) E numbers (c) Irradiation (d) Pasteurization

557. Multiple effect evaporation accounts for-----------
(a) Lower operating costs (b) Steam economy
(c) Investment economy (d) None of the above

558. What process should be carried out on vegetables before they are frozen?
(a) Chopping (b) Drying (c) Blanching (d) None of the above

559. At what temperature does blast freezing occur?
(a) -18^0C (b) -35^0C (c) -22^0C (d) -25^0C

560. What additive prevents foods with oil and water from separating?
(a) Antioxidant (b) Emulsifier (c) Chelating agent (d) All of the above

561. Which of the following is a form of fruit pretreatment?
(a) Sulfuring (b) Blanching (c) Syruping (d) All of the above

562. Function (s) of pretreatment of fruit----------
(a) To slow the action of enzymes (b) To prevent browning
(c) To increase nutrients (d) Both a and b

563. Which of the following is required to precipitate pectin in syrup?
(a) Water (b) Heat (c) Enzymes (d) None of the above

564. Freezer burn results from---------
(a) Chilling quickly (b) Airtight moisture resistant packaging
(c) Poor packaging resulting in moisture loss (d) Both a and b

565. The rate of heat transfer is a product of overall heat transfer co-efficient, the difference in temperature and the------------
(a) Heating volume (b) Heat transfer area (c) Nusselt number (d) None of the above

566. Most important factor to be considered in the selection of packings for absorbers is the --------- of packing.
(a) Porosity (b) Durability (c) Cost (d) Size

567. What are the labeling requirements for irradiated food?
(a) No label is required (b) A radura symbol with a treatment statement
(c) A radura symbol with a health warning (d) All of the above

568. An effective physical means to remove or destroy microorganisms is----------
(a) Heat (b) Filtration
(c) Ultraviolet radiation (d) All of the above

569. Chemical agents applied to tissue to prevent infection are called-----------
(a) Antiseptics (b) Disinfectants (c) Germicides (d) Herbicide

570. The ratio of change in volume to the original volume is called-----------
(a) Lateral strain (b) Poisson's ratio (c) Volumetric strain (d) Linear strain

571. The process by which a microbial population is reduced to safe levels is called--------
(a) Disinfection (b) Sterilization (c) Sanitization (d) Pasteurization

572. The movement of digestion products, electrolytes, vitamins, and water across the GI tract epithelium and into the underlying blood and lymphatic vessels is called------------
(a) Ingestion (b) Absorption (c) Digestion (d) Secretion

573. What is not true about the genetically modified 'Bt crops'?
(a) The technique has been applied to several species of crops
(b) The source organism for the gene is a soil bacterium
(c) The transferred gene exists in several allelic forms
(d) Bt brinjal has been commercially exploited

574. Loss of heat from untagged steam pipe to the ambient air is by-------------
(a) Convection (b) Radiation (c) Conduction (d) All of the above

575. Rate of adsorption increases as the-----------
(a) Pressure decreases (b) Temperature decreases
(c) Temperature increases (d) Size of adsorbent increases

576. Which of the following techniques would be most useful to identify and quantify the presence of a known impurity in a drug substance?
(a) NMR (b) MS (c) IR (d) HPLC

577. Buffers are important to human systems because they------------
(a) Stabilize the pH of a solution (b) Assist in chemical reactions
(c) Bring a solution to a pH of 9.8 (d) Increase the number of hydrogen ions

578. Polymers are------------
(a) Broken down by hydrolysis (b) Made by a dehydration process
(c) Macromolecules (d) All of the above

579. Most productive species among the different vitamin B_{12} producing Pseudomonas is-----------
(a) *P. denitrificans* (b) *P. ovalis* (c) *P. aureofaciens* (d) None of the above

580. Which of the following Ascomycetes are used for riboflavin production?
(a) *Eremnothecium ashbyii* (b) *Ashbya gossypii*
(c) *Candida guilliermondia* (d) Both a and b

581. Column efficiency is measured in terms of number of plates which is-----------
(a) Inversely related to the square of the peak width
(b) Directly related to the square of the peak width
(c) Inversely related to the cube root of the peak width
(d) Directly related to the cube of the peak width

582. Which can be used as a mobile phase in HPLC applications?
(a) Any compound with solubility in liquid
(b) Any compound with limited solubility in liquid
(c) Any compound with non-solubility in liquid
(d) Any of the above

583. Helium is generally preferred as carrier gas over nitrogen and hydrogen because------------
(a) It is inert
(b) It has a lower viscosity
(c) It doubles up as a party gas for balloons and funny voices
(d) All of the above

584. Protein-energy malnutrition is a condition that arises when------------
(a) Calories and protein intake are inadequate
(b) Protein and carbohydrate intakes are inadequate
(c) Proteins are not complete
(d) Protein intakes are not in balance

585. All organic substances contain which element?
(a) Oxygen (b) Bromine (c) Hydrogen (d) Carbon

586. Electrophoresis is related to which type of chromatographic process?
(a) Gas chromatography (b) Solid-state chromatography
(c) Thin-layer chromatography (d) High-performance liquid chromatography

587. Which of the following methods could be used to check the molecular weight of purified protein?
(a) SDS-PAGE only (b) Mass spectrometry only
(c) Analytical SEC only (d) All of the above

588. Sedentary people with elevated triglyceride levels who continue to eat fatty foods are likely to have high levels of-------------- in their bloodstream
(a) High-density lipoprotein (HDL) (b) Low-density lipoprotein (LDL)
(c) Very-low density lipoprotein (VLDL) (d) Chylomicrons

589. Butter contains-----------

(a) Milk fat, water, whey
(b) Milk fat, water
(c) Milk fat, milk solids
(d) Milk fat, water, milk solids

590. What is the inability to satisfy basic food needs due to lack of financial resources or other problems?

(a) Malnutrition (b) Famine (c) Hunger (d) Food insecurity

591. Which of the following has the lowest overall heat transfer co-efficient?

(a) Molten sodium (b) Air (c) Earth (d) Water

592. Malnutrition refers specifically to-----------

(a) Vitamin deficiency
(b) Any nutritional deficiency
(c) Failure to absorb nutrients
(d) All of the above

593. Fats have many more calories than carbohydrates because they---------------

(a) Contain many more potential acetyl groups to enter Krebs' cycle
(b) Are lipids
(c) Have more bonds from which to extract energy
(d) Produce much more ATP in glycolysis

594. Which coenzymes are reduced in both glycolysis and the Krebs' cycle?

(a) Manganese (b) Coenzyme A (c) Vitamin A (d) NAD

595. Which element below is considered a trace mineral?

(a) Phosphorus (b) Sodium (c) Potassium (d) Chromium

596. Which lipoprotein contains high levels of cholesterol?

(a) LDL (b) HDL (c) VLDL (d) IDL

597. In natural convection heat transfer, the correlating parameter is the------------

(a) Eckert number (b) Bond number (c) Graetz number (d) Grashoff number

598. Nutritionally incomplete protein lacks----------

(a) One or more of the major vitamins
(b) One or more of the coenzymes
(c) Actin and myosin
(d) One of the essential amino acids needed by the body

599. Which of the following statements is untrue about protein secondary structure?

(a) The alpha helix, beta pleated sheet and beta turns are examples of protein secondary structure
(b) The ability of peptide bonds to form intramolecular hydrogen bonds is important to secondary structure
(c) The steric influence of amino acid residues is important to secondary structure
(d) The hydrophilic/hydrophobic character of amino acid residues is important to secondary structure

600. What is the key factor that determines protein quality?
(a) Amino acid profile (b) DNA
(c) Synthesis (d) Denaturation

601. The function of *Streptococcus* during milk fermentation is to-----------
(a) Produce aerobic environment (b) Produce anaerobic environment
(c) Form lactic acid (d) All of the above

602. Which of the following elements occur in proteins but NOT in carbohydrates or lipids?
(a) Hydrogen (b) Carbon (c) Nitrogen (d) Oxygen

603. Which of the following is a kind of protein?
(a) Sucrose (b) Enzyme (c) Sodium chloride (d) Triglyceride

604. Normally, due to the holding of the chilled fish--------------
(a) *Pseudomonas* increase in numbers
(b) *Achromobacters* decrease
(c) Flavobacteria increase temporarily and then decrease
(d) All of the above

605. Sarcina sickness of beer is caused by-----------------
(a) *Saccharomyces cerevisiae*
(b) *Pedicoccus cerevisiae*
(c) *S. carlsbergensis*
(d) *Zygomonas anaeroium*

606. In dough, the starch is digested into sugar through----------
(a) Amylase (b) Maltase (c) Lactase (d) Protease

607. Beer may be produced by-----------
(a) Fermenting grape (b) Germinating barley
(c) Fermentation of rice (d) All of the above

608. ELISA is based on------------
(a) Antigen-antibody interaction (b) Antigen-protein interaction
(c) Lectin-antibody interaction (d) All of the above

609. Psychrometry deals with the properties of gas-vapor mixture. Humidity can be determined by the measurement of the -------------of a fibre.
(a) Thermal conductivity (b) Electrical resistance
(c) Strength (d) None of the above

610. PCR is used to---------
(a) Detect HIV in suspected AIDS patients
(b) Detect mutations in the genes in suspected cancer patients
(c) Diagnose many genetic disorders
(d) All of the above

611. Consumption of which one of the following foods can prevent the kind of blindness associated with vitamin 'A' deficiency?
(a) 'Flavr Savr'tomato (b) Canolla
(c) Golden rice (d) Bt-Brinjal

612. Which one of the following is a climacteric type of fruit?
(a) Banana (b) Citrus (c) Litchi (d) Grape

613. Which chemical is used for controlling sprouting of onions in storage?
(a) Maleic Hydrazide (MH) (b) Ethylene
(c) GA (d) All of the above

614. The strain energy stored in a spring, when subjected to maximum load, without suffering permanent distortion, is known as------------
(a) Impact energy (b) Proof resilience (c) Proof stress (d) Modulus of resilience

615. One which is known as a non renewable energy is----------
(a) Wind (b) Biomass (c) Coal (d) Tides

616. Separation of two or more components of a liquid solution cannot be achieved by-----------
(a) Absorption (b) Evaporation
(c) Liquid extraction (d) Fractional crystallization

617. This type of fat lowers low density lipoprotein-------------
(a) Polyunsaturated (b) Saturated fat (c) Transfat (d) All of the above

618. All EXCEPT which of the following are correct about food irradiation?
(a) That it imparts radioactivity to food (b) That it is used in the United States
(c) That it eliminates harmful bacteria (d) That it used gamma rays

619. All EXCEPT which of the following are correct about pasteurization?
(a) That it is a lengthy process that destroys the nutrients in milk
(b) That it was developed in 1862 by Louis Pasteur
(c) That it is important in the dairy and juice industry
(d) That it is a sterilization process that destroys microorganisms

620. Foods that are labeled organic must comply with all EXCEPT which of the following?
(a) If *labeled made with organic ingredients*, the product must have at least 70% organic ingredients
(b) Must be grown on an organic farm
(c) Only foods that are "100% organic" or "organic" can display the USDA organic seal
(d) Foods are grown without toxic or persistent fertilizers

621. Which of the following factors is not an issue with food safety?
(a) Temperature
(b) Humidity
(c) Whether the ingredients were fresh or canned
(d) Acidity

622. Since nitrosamines have been found to be carcinogenic in animals, the FDA requires that all foods with nitrites must also contain ------------

(a) Sulfites (b) EDTA (c) Calcium propionate (d) Antioxidants

623. Substances found on the GRAS list are thought to be--------------

(a) Safe for adults but not for children in high amounts
(b) Unsafe for pregnant women
(c) Unsafe for human consumption
(d) Safe for human consumption

624. All EXCEPT which of the following is a red flag in terms of food safety?

(a) Thawing poultry on the countertop
(b) Eating moldy yogurt with mold removed
(c) Disposing of refrigerated hard boiled eggs within a week if not eaten
(d) Eating a grilled hamburger that has reached an internal temperature of 140^0C

625. Which of the following is least likely to cause food borne illness?

(a) Bacteria (b) Animal feces (c) Fungi (d) Soil and manure

626. Which of the following additives can cause adverse asthmatic reactions in sensitive people?

(a) BHT (b) Nitrates (c) Propyl gallate (d) Sulfites

627. Marasmus results from all EXCEPT which of the following?

(a) Inadequate fat intake
(b) Inadequate carbohydrate
(c) Over consumption of foods with pesticides
(d) Inadequate protein

628. Which of the following causes the majority of food infections in the United States?

(a) *Norovirus* (b) *Salmonella* (c) *Escherichia coli* (d) *Staphylococcus*

629. A protein having both structural and enzymatic traits is-----------

(a) Collagen (b) Trypsin (c) Myosin (d) Actin

630. Which of the following is not a primary function of protein?

(a) Growth and maintenance of cells
(b) Production of antibodies
(c) Provides good and readily available source of energy
(d) Tissue and nerve development

631. The terms "chewy", "fibrous", "gritty", "mealy", and "sticky" are important in---------- of foods.

(a) Grading (b) Flavor (c) Nutritional value (d) Texture

632. -------------is the process of thawing frozen products.

(a) Flaking (b) Tamping (c) Tempering (d) Blanching

633. -----------is the science of evaluating a food product for smell, appearance, taste and texture.
(a) Proximate analysis (b) Food chemistry
(c) Rheology (d) Sensory evaluation

634. When vegetables are stored they undergo a process termed ------------ that yields principally water and carbon dioxide.
(a) Respiration (b) Perspiration (c) Dehydration (d) Oxidation

635. Km value of enzyme is substrate concentration at ------------
(a) 1/2 V max (b) 2 V max (c) 1/4 V max (d) 4 V max

636. Which of the following foods would be most likely to allow the growth of *Clostridium botulinum*?
(a) Eggs (b) Canned grapefruit (c) Pickles (d) Tomato sauce

637. All EXCEPT which of the following contribute to malnutrition?
(a) Food security (b) Famine (c) Poor quality diets (d) Over consumption

638. Which of the following is added to fat-based products to keep them from going rancid?
(a) Propionic acid (b) Ascorbic acid (c) Vitamin K (d) Vitamin E

639. Which of the following can cause headaches and difficulty breathing?
(a) Calcium propionate (b) MSG
(c) Vitamin E (d) BHA

640. Which types of fish tend to contain the highest levels of mercury?
(a) Pollock, sole, shrimp (b) Tilefish, swordfish, shark
(c) Tuna, salmon, cod (d) Mussels, scallops, lobster

641. Heavy, thick, slimy film of bacteria in the slow process of vinegar manufacture reduces-----
(a) The rate of hydrogenation (b) The rate of acetification
(c) The rate of esterification (d) The rate of purification

642. The potential spoilage organism in beer is---------------
(a) *Saccharomyces diastaticus* (b) *S. carlsbergensis*
(c) *S. cerevisiae* (d) *S. lactis*

643. Which nutrient is added to some foods to help prevent osteoporosis?
(a) Vitamin E (b) Vitamin C (c) Calcium (d) Niacin

644. Which of the following long-term complications can result from food poisoning?
(a) Rheumatoid arthritis (b) Kidney disease
(c) Nerve damage (d) All of the above

645. The first symptoms of food poisoning can occur------------
(a) Immediately (b) Within two to 48 hours after eating
(c) From two days to a week after eating (d) Any of the above

646. Which of these groups has a greater risk of getting food poisoning than the others?
(a) Smokers
(b) Heavy antacid users
(c) People who drink alcohol at least once a week
(d) They all have an equally high risk

647. What's an easy way to reduce the amount of potentially cancer-causing heterocyclic amines (HCAs) that form when chicken is broiled or grilled?
(a) Keep the chicken refrigerated until just before cooking
(b) Remove the skin after the chicken is cooked
(c) Remove the skin before the chicken is cooked
(d) Before broiling or grilling the chicken, pre-cook it in a microwave for a few minutes and pour off the juices

648. Eighty percent of all food poisoning from meat and poultry is caused by-------------
(a) *E. coli* O157:H7 (b) *Salmonella and Campylobacter*
(c) *Staphylococcus* (d) *Clostridium*

649. How can you tell if an egg is contaminated with *Salmonella*?
(a) The shell is cracked (b) The shell has dried chicken feces on it
(c) The egg hasn't been kept refrigerated (d) None of the above

650. A roasted chicken is thoroughly cooked when---------------
(a) The thigh reaches an internal temperature of 180°F
(b) The juices run clear
(c) The leg moves easily in its socket
(d) Any of the above

651. Which fish are least likely to contain chemical contaminants like PCBs, mercury, and pesticides?
(a) Lean ocean fish like cod, flounder, and haddock
(b) Freshwater fish caught in inland lakes like bluefish and lake trout
(c) Farm-raised catfish
(d) Canned tuna

652. Which of the following occurs during the coffee roasting period?
(a) Sugars convert to starch (b) Caffeol is released
(c) Caffeine content increases (d) All of the above

653. More than 90 percent of seafood poisoning cases would be eliminated if people----------
(a) Bought only government-inspected seafood
(b) Cooked their shellfish thoroughly
(c) Made sure their seafood was kept on ice until right before it was cooked
(d) Only ate fish they caught themselves

654. Catabolic processes-----------
(a) Make complex molecules from simpler ones
(b) Break complex molecules into simpler ones
(c) Occur only in autotrophs
(d) Occur only in heterotrophs

655. Which fluid does not experience shearing stress during flow?
(a) Pseudoplastic (b) Dilatant (c) Newtonian (d) Inviscid

656. A culture system with constant environmental conditions maintained through continual provision of nutrient and removal of wastes is called -------------- culture system.
(a) Continuous (b) Batch (c) Fed-batch (d) Semi continuous

657. Which of the following is considered as a disadvantage to the adsorption method of immobilization?
(a) It is possible to separate and purify the enzymes while being immobilized
(b) The enzymes are not usually deactivated by adsorption
(c) The adsorption is a reversible process
(d) State of immobilization is very sensitive to solution pH, ionic strength and temperature

658. Beta amylase is------------
(a) Endoenzyme (b) Exoenzyme
(c) Saccharifying enzyme (d) Both b and c

659. Enzymes act as antimicrobials--------------
(a) By depriving an organism of a necessary metabolite
(b) By generating a substances toxic to the organism
(c) By attracting a cell wall component
(d) All of the above

660. The bitter taste of the high protein materials is reduced by using------------
(a) Invertase (b) Pectinase (c) Protease (d) None of the above

661. Oxidizing which of the following substances yields the most energy?
(a) Proteins (b) Glucose (c) Fatty acids (d) Alcohol

662. Enzymes having slightly different molecules structure but performing identical activity are--
(a) Apoenzymes (b) Isoenzymes (c) Holoenzymes (d) Coenzymes

663. Aflatoxins are potent carcinogens in-------------
(a) The brain (b) The liver (c) The lungs (d) None of the above

664. Which one of the following statements is incorrect?
(a) SFAs with 14 and 16 fatty acids are strongly associated with raised LDL levels
(b) Replacement of SFAs with n-6 PUFAs reduces plasma LDL levels
(c) For optimal plasma lipid profiles, n-6 PUFAs should contribute at least 10% of dietary energy
(d) None of the above

665. The first food designated as a Food for Specified Health Use (FOSHU) in Japan was---------
(a) An oat bran-enriched breakfast cereal (b) A plant sterol-enriched margarine
(c) An isotonic sports drink (d) A hypoallergenic rice grain

666. Which one of the following is not part of the usual definition for a functional food?
(a) It is consumed as part of a normal food pattern
(b) It is not a pill, a capsule or any form of dietary supplement

(c) It has physiological benefits and/or reduces the risk of chronic disease beyond basic nutritional requirements
(d) None of the above

667. Lignans are part of which family of compounds?
(a) Carotenoids (b) Polyphenols (c) Phytosterols (d) None of the above

668. As part of their molecular structures, hydrocarbon carotenoids contain---------
(a) No hydroxyl groups (b) 1 hydroxyl group
(c) 2 hydroxyl group (d) >2 hydroxyl groups

669. The major organism used in the microbial production of citric acid is-------------
(a) *Penicillin notatum* (b) *Rhizopus nigrificans*
(c) *Aspergillus niger* (d) *Lactobacillis delbrueckii*

670. The aerated submerged bacterial fermentation to produce vitamin B_{12} employs strains of----
(a) *Propionibacterium* (b) *Pseudomonas*
(c) *Acetobacter* (d) Both a and b

671. What role does O_2 play in aerobic respiration?
(a) It plays no role
(b) It combines with acetyl-CoA at the start of the Krebs cycle
(c) It is given off as a by-product during the oxidation of pyruvate
(d) It is the final electron acceptor at the end of the electron transport chain

672. In which of the following organisms enzymes are absent--------------
(a) Fungi (b) Viruses (c) Algae (d) Bacteria

673. Enzymes exist in the cells as-------------
(a) Solution (b) Crystals (c) Solids (d) Colloids

674. What is the purpose of aging wine and some distilled spirits in oak barrels?
(a) Aging in oak barrels decreases the acidity and removes undesirable impurities
(b) Aging imparts the essence of oak, a common flavor noted in all alcoholic beverages
(c) Oak barrels are the only storage container that protects the beverages from bacterial contamination
(d) Putting spirits in oak barrels prevents the alcohol from aging any further after the distillation is complete

675. During aerobic respiration, $FADH_2$ is produced in---------
(a) Glycolysis (b) The oxidation of pyruvate
(c) The Krebs cycle (d) Fermentation

676. Taste buds are located primarily on the-------------
(a) Upper palate (b) Tongue (c) Gums (d) Turbinates of the nose

677. The types of tastes are---------------
(a) Hot, bland, sweet and sour (b) Gradations between sweet and bitter
(c) Bitter, sour, sweet and salty (d) All of the above

678. Organisms that do not have the ability to produce or synthesize their own food are called----

(a) Anaerobic (b) Autotrophs (c) Exergonic (d) Heterotrophs

679. Proteins and fats can be nutritional sources of energy provided that------------

(a) They are converted into glucose

(b) The enter their own pathways that are separate from the glucose metabolic pathways

(c) They are degraded completely into atoms before entering a pathway

(d) They are modified so that they can enter the glucose metabolic pathways

680. The symptoms of ciguatera, the most common finfish poisoning in the U.S., are---------

(a) Headache and vomiting

(b) Blood in the stool

(c) Nausea, cramps, and diarrhea, followed by a "pins and needles" sensation around and inside the mouth

(d) Any of the above

681. What is the advantage of adding hops to the wort during the beer making process?

(a) Adds a desirable bitterness to the beer (b) For antibacterial action

(c) Helps retain the beer's foamy head longer (d) All of the above

682. Which warnings does the Food and Drug Administration require on the labels of foods that contain the fake fat olestra (Olean)?

(a) May cause abdominal cramping (b) May cause loose stools

(c) Inhibits the absorption of some vitamins (d) All of the above

683. What are 2 of the HACCP Principles in correct order?

(a) Conduct a food safety hazard analysis, and Critical Limits

(b) Establish effective record keeping, and Identify the CCP's, including their location, procedure and process

(c) Identify the CCP's, including their location, procedure and process, and establish procedures for verification

(d) Establish corrective actions, and Monitoring

684. What are the three types of food hygiene hazards?

(a) Animal hair, Pesticides and Salmonella (b) Physical, Chemical, Biological

(c) Chemical and Biological (d) Microbiological, Physical and Chemical

685. Naringinase, an enzyme commonly used for removal of bitter taste from citrus juice is obtained from------------

(a) *A. niger* (b) *A. flavus* (c) *A. oryzae* (d) *S. cerevisiae*

686. Protease, which is used for flavouring of sake and haze removal in sake is produced by-----

(a) *A. oryzae* (b) *A. flavus* (c) *B. cereus* (d) *A. niger*

687. What is an example of a biological hazard?

(a) Salmonella (b) Dirt (c) Cleaners (d) Antibiotics

688. What does WHS stand for?
(a) Well being Health and Safety (b) World Health and Safety
(c) Work Health and Safety (d) Work Help and Safety

689. What is the HACCP system?
(a) Physical, Chemical and Biological Hazards
(b) Accurately monitoring food hygiene hazards
(c) Identifying the CCP's, including their location, procedure and process
(d) A systematic analysis of all steps and regular monitoring of the control points

690. Agarose extracted from sea weeds finds use in------------
(a) Gel electrophoresis (b) Spectrophotometry
(c) Tissue culture (d) PCR

691. Continuous addition of sugars in 'fed batch' fermentation is done to----------
(a) Degrade sewage (b) Produce methane (c) Obtain antibiotics (d) Purify enzymes

692. DNA is found in------------
(a) Chromosomes (b) Plastids (c) Mitochondria (d) All of the above

693. Genetic engineering is-------------
(a) Plastic surgery (b) Study of extra nuclear genes
(c) Addition or removal of genes (d) All of the above

694. --------------stands for a system that is used to enhance food safety in food processing, packaging, storage, distribution, and preparation.
(a) Good manufacturing practices
(b) Hazard analysis and critical control point
(c) High accuracy and contamination control point
(d) Best management practices

695. -------------are places in the food processing system where the lack of proper control can result in a safety risk for the consumer.
(a) Concentrated contamination control processes
(b) Contamination processing points
(c) Critical contamination places
(d) Critical control points

696. A food additive that promotes or produces a desired physical state or texture is called as-----
(a) Enzyme (b) Formulation aid (c) Firming agent (d) Humectants

697. A bacteria that can contaminate poultry products and cause food borne illness in humans is ---------
(a) Lactobacillus (b) Clostridium (c) Gram Positive (d) Salmonella

698. Two factors that accelerate rancidity in food products are------------
(a) Temperature and light (b) Light and oxygen
(c) Light and moisture (d) Light and soluble minerals

699. Which of the following organism is found at the initial stages in the batter of *Idli*?
(a) *Leuconostoc mesentroides* (b) *Aspergillus niger*
(c) *Bacillus natio* (d) None of the above

700. The most effective way to eliminate living microorganisms in spices is ----------
(a) Freezing (b) Irradiation (c) Heat (d) Selected chemicals

701. An addition of a nutrient to foods such as adding vitamin A to milk is called as-------------
(a) Pasteurization (b) Fermentation (c) Nutrification (d) Fortification

702. Only lactic acid bacteria can ferment sugars and nutrients in pickles because they ------------
(a) Use a natural occurring enzyme (b) Are tolerant to salt levels
(c) Produce lactic acid (d) Use acetic acid

703. A list of ingredients must be included on a food label. The first ingredient listed is by its amount of-------------
(a) Percent protein (b) Grams of carbohydrates
(c) Total weight (d) Fat content

704. Genomics is the study of------------
(a) Genes in general (b) Human genes
(c) Genomes in general (d) Human genome

705. Which of the following statements about column chromatography is correct?
(a) Affinity chromatography involves the attachment to the column matrix of ionic groups known that bind to the wanted protein
(b) In reverse phase chromatography the wanted protein can be selectively eluted by solutions of different hydrophobicities or ionic strengths
(c) Ion-exchange chromatography separates proteins according to their size
(d) Gel-filtration chromatography separates proteins on their ability to bind to specific groups on the column matrix

706. Which one of the following processes involves southern blotting technique?
(a) rDNA technology (b) DNA finger printing
(c) Production of transgenic plants (d) Formation of suspension culture

707. When energy-depleted elements associated with a proton are accepted by an organic molecule, the process is called------------
(a) Fermentation (b) Anaerobic (c) Aerobic (d) Catabolism

708. The end product of glycolysis is-----------
(a) NADH (b) Acetyl-CoA (c) Lactate (d) Pyruvate

709. Which one of the following physiologically important classes of metabolites is produced by bacterial fermentation of non-digestible carbohydrates in the large intestine?
(a) Short-chain fatty acids (b) Long-chain fatty acids
(c) Bile acids (d) Amino acids

710. Which of the following foods has the highest glycemic index (GI)?
(a) Apples (b) Beans (c) Bananas (d) Milk

711. Which one of the following functions is not carried out by lipids?
(a) Essential component of cell membranes (b) Insulation to maintain body temperature
(c) Hormone precursors (d) Vehicle for absorption of water-soluble vitamins

712. Which one of the following is a monounsaturated fatty acid?
(a) Butanoic acid (b) Hexadecenoic acid (c) Octadecanoic acid (d) Eicosapentaenoic acid

713. The difference between *cis-* and *trans-* fatty acids is----------
(a) The number of double bonds (b) The length of their carbon chains
(c) The position of the first double bond (d) The configuration around the double bond

714. *Trans*-fatty acids are undesirable in the diet because they----------
(a) Raise LDL cholesterol
(b) Lower HDL cholesterol
(c) Pack closer together than *cis*-fatty acids in cell membranes, resembling saturated fatty acids
(d) All of the above

715. Triglycerides make up roughly what proportion of total lipids in the average diet-------
(a) 35% (b) 55% (c) 75% (d) 95%

716. Which one of the following statements is correct?
(a) The major fatty acid in olive oil is stearic acid (18: 0)
(b) The major fatty acid in coconut oil and palm oil is butyric acid (4: 0)
(c) Triglycerides are good emulsifiers, whereas phospholipids are not
(d) Eicosanoids are derived from 20-carbon polyunsaturated fatty acids

717. The principle function of low-density lipoproteins is to-----------
(a) Transport cholesterol from the small intestine to the liver
(b) Transport cholesterol from the liver to the tissues
(c) Transport cholesterol and other lipids from the tissues to the liver
(d) None of the above

718. Which one of the following statements is incorrect?
(a) Fat utilisation is stimulated by noradrenaline
(b) Insulin resistance is a feature of obesity
(c) Fat deposited within the chest cavity and abdomen is associated with increased health risks
(d) None of the above

719. Which part of its chemical structure differentiates one amino acid from another?
(a) The functional group (b) The carboxyl group
(c) The amino group (d) None of the above

720. The usefulness of fermentation as a means of deriving energy is limited because---------
(a) It cannot generate enough ATP
(b) It produces too much NH_2
(c) The end products are toxic to the producer
(d) It uses more energy than it produces

721. Which of the following is not a product of fermentation?
(a) CO_2 (b) O_2 (c) Ethanol (d) All of the above

722. Specificity of protein in enzyme action depends upon-------------
(a) Active sites (b) Km constant
(c) Linear sequence of amino acids (d) Turn over number

723. ELISA tests are used to-------------
(a) Isolate DNA sequence (b) Purity proteins
(c) Separate viral RNA (d) Identify specific proteins

724. In order to make white wine from red grapes,-------------- must be removed from the juice before fermentation.
(a) Glucose (b) The seeds (c) Ascorbic acid (d) The skins

725. Enzymes functional inside the cells are called-------------
(a) Endoenzymes (b) Exoenzymes (c) Apoenzymes (d) Isoenzymes

726. Which of the following is NOT a process in the making of champagne?
(a) Fortification (b) Riddling (c) Disgourgement (d) Both a and c

727. ---------------is produced by using a yeast that settles to the bottom of the tank.
(a) Ale (b) Malt (c) Lager beer (d) None of the above

728. Which of the following is most efficiently absorbed?
(a) Fat (b) Protein (c) Carbohydrates (d) Alcohol

729. Which one of the following carbohydrates would need to be malted during the fermentation process of making beer?
(a) A monosaccharide such as glucose (b) A disaccharide such as maltose
(c) A polysaccharide such as starch (d) A yeast to start the fermentation process

730. Alcohol is most damaging to------------
(a) The brain cells since alcohol can be used as an energy source even before glucose
(b) The stomach cells since this is the site where ADH (alcohol dehydrogenase) metabolizes alcohol
(c) The cells of the gastrointestinal tract since they are in direct contact with ingested alcohol
(d) The liver cells where alcohol is oxidized

731. A varietal wine-------------
(a) Is made with a variety of grapes
(b) Is always an American wine

(c) Will go well with many different dishes
(d) Is named for the principle grape variety with which it is made

732. Most red wines should be served at room temperature because-------------
(a) They will have clearer color (b) Their bouquet can be better appreciated
(c) They taste better with hot food (d) It's a traditional custom

733. The reason for storing wine bottles on their sides is---------------
(a) To keep the cork moist (b) To save storage space
(c) Because the wine ages better this way (d) Because the bottles look nice in a wine rack

734. The reason some cheeses are softer than others is because------------
(a) The milk was fresher
(b) An enzyme in the cheese prevented it from getting hard
(c) Less of the whey was extracted from the curd
(d) The cheese was ripened in a warm place

735. Genetic engineering manipulates gene products at the level of the-----------
(a) Protein (b) Amino acid (c) DNA (d) RNA

736. The little bumps you can see on someone's tongue are------------
(a) Taste buds (b) Papillae (c) Gustatory pimples (d) Latin papula

737. The bubbles in champagne are due to--------------
(a) The kind of cork used as a stopper (b) The special variety of grapes used
(c) Added carbon dioxide (d) A second fermentation after the wine is bottled

738. Which of the following are the six steps in wine-making?
(a) Crushing, fermenting, filtering, fining, maturing, bottling, and resting
(b) Crushing, fermenting, maturing, resting, bottling, and barreling
(c) Crushing, fermenting, filtering, fining, maturing, and bottling
(d) Crushing, fermenting, racking, maturing, filtering, and bottling

739. Which of the following most accurately describes the role of DNA for our cells?
(a) DNA physically moves parts around a cell to determine the cell's function
(b) DNA includes instructions for building proteins
(c) DNA contains a code that allows it to turn into a protein
(d) DNA is an enzyme that assists in chemical reactions

740. The membranes of our cells are made mostly of------------
(a) Cholesterol (b) Proteins (c) Polysaccharides (d) Lipids

741. Which of the following methods of infection control requires the use of an autoclave---------
(a) Disinfection (b) Ultrasonic cleaning (c) Sterilization (d) Sanitization

742. Which of the following is permissible in ultrasonic cleaning?
(a) Place all instruments together in the special bath
(b) Place all instruments with hinges or ratchets in the ultrasonic cleaner in the closed position

(c) Remove the instruments from the ultrasonic cleaner and immediately wrap in a towel without rinsing
(d) Reuse the ultrasonic cleaning solution for several baths

743. The cleaning process that is required for all instruments that penetrate the skin or that come in contact with normally sterile areas of the tissue and internal organs is------------
(a) Sterilization (b) Sanitization (c) Disinfection (d) Ultrasonic cleaning

744. The polymerase chain reaction is a technique that------------
(a) Selectively replicates RNA (b) Probes and selectively replicates DNA
(c) Selectively replicates DNA (d) Probes DNA

745. A "probe" in molecular biology is-------------
(a) An instrument used to manipulate cells in culture
(b) A type of vector system
(c) A DNA or an RNA molecule used in hybridization reactions
(d) Probes are not used in molecular biology

746. In gel electrophoresis, DNA molecules migrate from ------------- to------------ ends of the gel.
(a) Negative to positive (b) Basic to acidic
(c) Long to short (d) Positive to negative

747. Which of the following techniques can be used to determine the defective gene and for developing cancer?
(a) Western blot (b) Southern blot (c) Northern blot (d) Eastern blot

748. Both DNA gel electrophoresis and SDS-PAGE of proteins are similar because-----------
(a) In both cases molecules migrate to the anode
(b) Both techniques rely on a constant charge to mass ratio
(c) Both techniques utilize the sieving properties of gels
(d) All of the above

749. X-rays cause-------------
(a) The formation of thymine dimers (b) Ionization of water in the cell
(c) Heat (d) None of the above

750. Why is golden rice pale yellow in color?
(a) It is rich in chlorophyll *a* (b) It is rich in beta-carotene
(c) It is rich in chlorophyll *b* (d) It is rich in phycobilins

751. Which of the following products would be MOST likely to benefit from a mass marketing approach?
(a) Health club (b) Chewing gum (c) Computers (d) Luxury cars

752. Eating foods that are low on the Glycemic Index will prevent which disease?
(a) Type 2 diabetes (b) Coronary Heart Disease
(c) Cancer (d) All of the above

753. The name of the process used to identify appropriate target markets is-----------

(a) Market segmentation (b) Niche marketing
(c) Traditional marketing (d) Mass marketing

754. An organic nutrient essential to an organism's metabolism that cannot be synthesized itself is termed a/an------------

(a) Trace element (b) Micronutrient (c) Growth factor (d) Essential nutrient

755. An organism that can synthesize all its required organic components from CO_2 using energy from the sun is a-----------

(a) Photoautotroph (b) Photoheterotroph (c) Chemoautotroph (d) Chemoheterotroph

756. The strong bonds that usually form between carbon and hydrogen are called ----------- bonds.

(a) Ionic (b) Hydrogen (c) Elemental (d) Covalent

757. Which nutrient plays key roles in conducting nerve impulses, muscular contraction, and the maintenance of water balance?

(a) Protein (b) Fats (c) Carbohydrates (d) Minerals

758. Which of the following is true about biscuits?

(a) The dough is kneaded
(b) The dough can re-rolled several times
(c) The biscuit cut outs are placed on a greased sheet
(d) Both a and c

759. The RDA refers to------------

(a) The weekly amounts of nutrients needed for healthy individuals
(b) The amount of food intake recommended for healthy weight loss
(c) The daily amounts of nutrients recommended for healthy individuals
(d) The daily amounts of nutrients recommended for all individuals

760. Recent studies indicate that coffee drinkers have a lower incidence of------------

(a) Multiple sclerosis (b) Cancer
(c) Parkinson's Disease (d) Alzheimer's disease

761. Vegetable oil is a good source of-------------

(a) Magnesium (b) Vitamin D (c) Vitamin E (d) Carbohydrate

762. What is the goal of lowing sodium intake?

(a) To lower blood cholesterol levels (b) To lower blood pressure
(c) To promote weight loss (d) To lower rates of colon cancer

763. Meats are an excellent source of------------

(a) Vitamin C (b) Calcium (c) Fiber (d) Iron

764. The incidence of dental caries is associated with------------

(a) Amount of carbohydrate consumed (b) Hygiene practices
(c) Host genetic factors (d) All of the above

765. *Salmonella* variants cause all the following EXCEPT--------------

(a) Salmonellosis (b) Typhoid fever (c) Gastroenteritis (d) Typhus

766. Identify the following type of hepatitis that can lead to cancer.

(a) Hepatitis A (b) Hepatitis B (c) Hepatitis D (d) Hepatitis E

767. Which of the following would be most readily killed by chlorination of water?

(a) Viruses (b) Protozoa (c) Cysts (d) Bacterial endospores

768. Which of the following statements about enzymes or their function is true?

(a) Enzymes do not alter the overall change in free energy for a reaction

(b) Enzymes are proteins whose three-dimensional form is key to their function

(c) Enzymes speed up reactions by lowering activation energy

(d) All of the above

769. Which of the following may NOT be used as a claim on a food label?

(a) Calorie free (b) Low calorie (c) Sugar free (d) Low sugar

770. Which of the following food processing operations is NOT for cooling food products?

(a) Air blast (b) Ice water bath (c) Extrusion (d) Vacuum oven

771. Which one of the following is a type of food preserved, in part, by bacteria?

(a) Yoghurt (b) Bread (c) Wine (d) Whole milk

772. Which of the following is NOT an essential function of a food container?

(a) Tamper-resistant (b) Refrigerator fit (c) Light protection (d) Sanitary protection

773. Which of the following work together to maintain chemical, fluid, and electrical balance between tissue cells and blood?

(a) Sodium and potassium (b) Calcium and phosphorus

(c) Iron and vitamin C (d) Calcium and vitamin D

774. A fatty acid does NOT contain which of the following elements?

(a) Carbon (b) Nitrogen (c) Oxygen (d) Hydrogen

775. The ------------ regulates genetically engineered microbes used in natural pesticides.

(a) Environmental Protection Agency

(b) United States Department of Agriculture

(c) United States Department of Genetic Engineering

(d) Food and Drug Administration

776. In which of the following foods solanine is considered as a toxin?

(a) Potato (b) Tomato (c) Coffee (d) Tea

777. Which one of the following is not a feature of marasmus?

(a) Oedema (b) Muscle wasting

(c) Increased susceptibility to infection (d) Shrivelled skin

778. Which one of the following is not a consequence of iron deficiency?
(a) Anaemia (b) Impaired night vision
(c) Poorer pregnancy outcome (d) Reduced work capacity

779. Deficiency of which nutrient is associated with increased risk of infection in children, growth failure throughout childhood and delayed sexual maturity?
(a) Iodine (b) Zinc (c) Riboflavin (d) Calcium

780. The recommended dietary allowance (RDA. for a nutrient is-------
(a) Estimated average requirement (EAR) plus one standard deviation
(b) Estimated average requirement (EAR) plus two standard deviations
(c) Adequate intake (AI) plus one standard deviation
(d) Tolerable upper intake level (UL) minus two standard deviations

781. For normal brain development, an appropriate ratio of which of the following is particularly important?
(a) Sodium and potassium (b) Calcium and phosphorus
(c) Essential and non-essential amino acids (d) n-6 and n-3 fatty acids

782. Lactoferrin in human milk-------------
(a) Promotes calcium absorption (b) Binds iron to restrict bacterial growth
(c) Facilitates fat digestion (d) None of the above

783. Low birth weight in infants is associated with which disease(s) in adulthood?
(a) Type 2 diabetes (b) Hypertension
(c) Coronary heart disease (d) All of the above

784. What are probiotics?
(a) Killed microorganisms that provide health benefits
(b) Live microorganisms that provide health benefits
(c) Bacterial toxins that provide health benefits
(d) Live microorganisms that could be harmful

785. Age-related loss of muscle mass, strength and function is called------------
(a) Neutropenia (b) Sarcopenia (c) Osteopenia (d) Thrombocytopenia

786. Deficiency of which one of these micronutrients affects taste sensitivity
(a) Copper (b) Selenium (c) Iodine (d) Zinc

787. Breakdown of carbohydrates in the colon results in the production of-------------
(a) Short-chain fatty acids (b) Medium-chain fatty acids
(c) Long-chain fatty acids (d) None of the above

788. Which of the following pairs of fatty acids are essential for production of prostaglandins and leukotrienes in the brain?
(a) Stearic and oleic acid (b) Linoleic and linolenic acid
(c) Arachidonic and docosahexaenoic acid (d) None of the above

789. An emulsifier--------------

(a) Prevents the separation of oil and water in food
(b) Maintains the shape or crispness of fruits and vegetables
(c) Controls insects and pests
(d) Produces or stimulates CO_2 production

790. A food additive that retards rancidity of unsaturated oils and prevents browning in fruits and vegetables that occur during exposure to oxygen is called an -----------

(a) Anti-caking free-flowing agent (b) Antimicrobial agent
(c) Antioxidant (d) Antibuffer agent

791. Most spoilage bacteria grow at-------------

(a) Acidic pH (b) Alkaline pH (c) Neutral pH (d) Any of the pH

792. Which of the following acid will have higher bacteriostatic effect at a given pH?

(a) Acetic acid (b) Tartaric acid (c) Citric acid (d) Maleic acid

793. Which compounds are great for bone health?

(a) Vitamin K and calcium (b) Potassium and magnesium
(c) Vitamin D and manganese (d) Vitamins A and B6

794. What compound promotes haemoglobin creation and metabolises proteins?

(a) Potassium (b) Fibre (c) Iron (d) Vitamin A

795. Which of these things won't green leafy vegetables help you with?

(a) Protect against disease (b) Boost energy levels
(c) Increase hair growth (d) Boost immune system

796. Leptin is a hormone produced by---------

(a) Stomach (b) Liver (c) Pancreas (d) Adipose tissue

797. The disaccharide sucrose consists of which two monosaccharides?

(a) Glucose and galactose (b) Glucose and fructose
(c) Fructose and galactose (d) None of the above

798. Which one of the following oligosaccharides contains α-linked sugars and is therefore digestible by humans?

(a) Maltodextrins (b) Raffinose (c) Inulin (d) Stachyose

799. Which one of the following statements is correct?

(a) Starch is a gluco-oligosaccharide, whereas cellulose is a fructo-oligosaccharide
(b) Amylase converts starch into maltose and cellulose to cellobiose
(c) Starch and cellulose both contain chains of glucose molecules mainly in α-1,4 linkage
(d) None of the above

800. Which one of these statements about starch digestion is incorrect?

(a) It begins in the mouth
(b) It occurs in the stomach

(c) It requires the action of pancreatic enzymes

(d) It requires brush border enzymes for completion

801. The major product of amylose digestion by the enzyme amylase is----------

(a) Glucose (b) Fructose

(c) Maltose and maltotriose (d) Oligosaccharides

802. Fermentation in which two or more than two microorganisms are utilized is called-----------

(a) Continuous fermentation (b) Anaerobic fermentation

(c) Dual fermentation (d) All of the the above

803. The lowest temperature that kills all microorganisms in a liquid suspension in 10 minutes is known as---------------

(a) Thermal death point (b) Thermal death temperature

(c) F value (d) D value

804. D value in death rate curve indicates time required for survival or death rate curve to traverses thorugh--------------

(a) One log cycle (b) 1/10 log cycle (c) 1/100 log cycle (d) None of the above

805. In the spoilage of meat, changes in fat occur due to-----------

(a) Actinomycete (b) Luminous bacteria (c) Achromobacter (d) Proteolytic bacteria

806. Which of the following disinfectants acts by disrupting microbial membranes?

(a) Cationic detergents (b) Halogens

(c) Organic acids (d) Aldehydes

807. Which of the following prevent the germination of *Clostridium botulinum*?

(a) Sorbic acid (b) Benzoic acid (c) Nitrate and nitrites (d) SO_2

808. The rice beer or wine obtains from *Aspergillus oryzae* is known as----------

(a) Sake (b) Sherry (c) Dry wine (d) None of the the above

809. Ozone has a strong microbicidal action against------------

(a) Bacteria (b) Fungi

(c) Parasites and viruses (d) All of the above

810. Sausages are a class of chopped or ground meat products that have under gone-------------

(a) Fermentation (b) Aging (c) Drying (d) All of the above

811. Which of the following is an example of a food borne disease-caused by parasite?

(a) *Canpylobacter jcjuni* (b) *Salmonella enteritidus*

(c) *Clostridium perfringens* (d) *Trichinella spiralis*

812. In recent years, the potential for food contamination has been increased due to-------------

(a) Multiple handling of food in supply chain from production to final customer

(b) Cross contamination of food

(c) Processing equipments

(d) Lack of federal, state and local government regulation of food products

813. A culture started with 4 cells and ended with 128 cells. How many generations did the cell go through?
(a) 64 (b) 32 (c) 6 (d) 5

814. Which of the following organisms have wide range of pH for growth?
(a) Fungi (b) Yeast (c) Bacteria (d) Viruses

815. Which of the following was the first widely used antiseptic and disinfectant?
(a) Chlorine (b) Phenol (c) Iodine (d) Alcohol

816. *Listeria monocytogenes* is --------------
(a) Salt tolerant (b) Can grow at refrigeration temperature
(c) Can grow better at lower O2 level (d) All of the above

817. T-2 is a trichothecene mycitoxin is synthesised by----------------
(a) *Alternaria spp.* (b) *Penicillium spp.* (c) *Fusarium spp.* (d) All of the above

818. Iodine sanitizers causes death of pathogens, due to-------------
(a) Destruction of mitochondria (b) Damage to DNA
(c) Cell wall damage (d) All of the above

819. Dry red chilli is adulterated with---------------
(a) Rhodamine B colour (b) Indigo carmine
(c) Metanil red (d) Both a and b

820. Mass spectrometers separate isotopes of different elements based on their------------
(a) Mass (b) Electric charge
(c) Mass by electric charge (d) None of the above

821. The Codex Alimentarius covers foods like-------------
(a) Processed (b) Raw food (c) Semi-processed (d) All of the above

822. The ultraviolet radiation (UVR) had range of electromagnetic spectrum-------------
(a) 10–400 nm (b) 600–800 nm (c) <10 (d) > 800 nm

823. MPEDA was set up to undertake the promotional work relating to the export of-------------
(a) Milk products (b) Meat products (c) Marine products (d) All of the above

824. International Food Standard (IFS) is designed for companies who undertake the -------- activities.
(a) Processing (b) Handling of loose food products
(c) Primary packing activities (d) All of the above

825. The anatomy of bar code contains-------------
(a) Number system character (b) Manufacturer code
(c) Product code and Check digit (d) All of the above

826. ISO 22000 standards are associated with-----------
(a) Safety and quality (b) Quality
(c) Safety (d) None of the above

827. The Global Food Safety Initiative (GFSI) is a private organization under--------------
(a) Belgian law (b) Denmark law (c) Bangladesh (d) England

828. Traceable supply chains from farm to end-user has become an important part of the overall food quality assurance system for consistency in--------------
(a) Top quality (b) Rebuild public confidence
(c) Safe and nutritious foods (d) All of the above

829. Supply chain management (SCM), represents----------------
(a) The management of the entire set of production
(b) Manufacturing/transformations
(c) Distribution and marketing activities
(d) All of the above

830. Each consignment of rice exported to U.S. shall be accompanied by-------------
(a) Phytosanitary Certificate
(b) The consignment is processed by registered rice milling
(c) A certificate of origin
(d) All of the above

831. The adulteration of honey with glucose is detected by---------------
(a) Fiehe's Test (b) Aniline chloride Test
(c) Both a and b (d) None of the above

832. Which of the following is the simplest functional food?
(a) Margarine with sterol extract (b) Oat meal
(c) Orange juice with calcium (d) Chocolates

833. Functional foods include-----
(a) Whole foods (b) Modified foods (c) Fortified foods (d) All of the above

834. The difference between organic food and non-organic food lies in-------------
(a) Fertilizing the crops (b) Weeding
(c) Managing pests and diseases (d) All of the above

835. What are the benefits of organic foods?
(a) They are often fresher (b) They have fewer pesticides and chemicals
(c) They are environment friendly (d) All of the above

836. What's the best way to tell the difference between organic and nonorganic foods?
(a) Give them the "squeeze test (b) Organic foods always cost more
(c) Look for the organic label (d) There is no difference

837. What portion of a food must be organic to allow a food maker to use the label "made with organic ingredients"?
(a) 0 to 10% (b) 10 to 25% (c) 25 to 70% (d) 70 to 95%

838. To carry the USDA organic label, produce cannot have--------------
(a) Manmade pesticides (b) Genetic engineering
(c) Irradiation (d) All of the above

839. The organic label lets you know that organic foods are-----------------
(a) Healthier (b) Better tasting (c) Both a and b (d) Neither a nor b

840. Why do organic foods often cost more than nonorganic foods?
(a) Higher taxes (b) Production costs (c) More organic farms (d) None of the above

841. Eating nonorganic produce can expose you to--------------
(a) Potentially harmful pesticide levels (b) Safe levels of pesticides
(c) No proven health problems (d) None of the above

842. Besides buying organic, you can limit exposure to pesticides by-------------
(a) Washing all uncooked fruits and vegetables
(b) Peeling and trimming fruits and vegetables
(c) Eating a variety of foods
(d) All of the above

843. Calcium bioavailability is reduced by------------
(a) Dietary fibre (b) Oxalates (c) Phytic acid (d) All of the above

844. Calcium is required for which of the following physiological processes?
(a) Blood clotting (b) Muscle contraction
(c) Nerve impulse transmission (d) All of the above

845. Which mineral is a structural component of thiamin, biotin, coenzyme A, acyl carrier protein and glutathione?
(a) Magnesium (b) Sulphur (c) Sodium (d) Calcium

846. A key storage form of iron is----------
(a) Haemoglobin (b) Transferrin (c) Myoglobin (d) Ferritin

847. Tea is a major dietary source of which one of the following trace elements?
(a) Manganese (b) Zinc (c) Copper (d) Selenium

848. The enzyme transketolase, part of the pentose phosphate pathway, requires which B-vitamin for optimal function?
(a) Biotin (b) Pantothenic acid (c) Folate (d) Thiamin

849. The release of glucose from glycogen stores in the liver and muscle requires which B-vitamin?
(a) Pantothenic acid (b) Thiamin (c) Pyridoxine (d) None of the above

850. Which two vitamins share a role in the conversion of homocysteine to methionine?
(a) Vitamin C and vitamin B_6 (b) Folate and vitamin B_{12}
(c) Niacin and biotin (d) None of the above

851. Approximately what percentage of the body's iron content is present in haemoglobin?
(a) 20% (b) 40% (c) 60% (d) 80%

852. Which one of the following food components is a promoter of iron absorption?
(a) Phytate (b) Oxalates (c) Polyphenols (d) Lactic acid

853. Which one of the following statements is incorrect?
(a) Catalase, which removes hydrogen peroxide, is a zinc metalloenzyme
(b) Vitamin E is the major antioxidant in the lipid environment of the cell, whereas vitamin C is active in the aqueous environment
(c) Carotenoids are important antioxidants because of their ability to quench singlet oxygen
(d) Iron and copper are kept in the bound state, generally attached to proteins, because they are potential catalysts for free radical formation

854. The Code of Criminal Procedure (CrPC) is the main legislation on procedure for administration of substantive criminal law in India. It was came into force on -----------
(a) 1st April 1974 (b) 1st April 1973 (c) 1st April 1975 (d) 1st April 1976

855. The objective of ISO-9000 family of quality management is------------
(a) Customer satisfaction (b) Employee satisfaction
(c) Skill enhancement (d) Environmental issues

856. Total Quality Management (TQM) focuses on-----------
(a) Employee (b) Customer (c) Both a and b (d) None of the above

857. While setting quality objective, ------------ to be considered.
(a) Material quality (b) Customer need (c) Market demand (d) All of the above

858. Which of the following is for environment management?
(a) ISO-9000 (b) ISO-14000 (c) ISO-26000 (d) ISO-31000

859. Which of the following was not given as a key feature of Total Quality Management?
(a) Using quality tools and techniques (b) Continuous improvement
(c) Identifying customers and their needs (d) Establishing clear specifications

860. A fundamental attribute of TQM is ------------
(a) Drawing control charts (b) Meeting ISO 9000 audit
(c) Top management's direct involvement (d) All of the above

861. Which of the following individuals may not benefit from dietary supplementation?
(a) Pregnant teenagers (b) Vegans
(c) Adults eating a well-balanced diet (d) Newborns

862. Which vitamin/mineral supplement has the potential risks of stomach pain, nausea, vomiting, and diarrhea?
(a) Niacin (b) Vitamin B6 (c) Selenium (d) Vitamin A

863. Which of the following is true concerning dietary supplements?
(a) There are specific rules to limit the serving size or amount of a nutrient in a dietary supplement.
(b) Dietary supplements need approval from the FDA before they are marketed.
(c) Supplement companies always have to provide the FDA with evidence that their supplement is safe.
(d) Dietary supplements are categorized within the general group of foods, not drugs.

864. What form can one find dietary supplements?
(a) Pills (b) Capsules (c) Powders (d) All of the above

865. By keeping food cold the growth of microorganisms is-------------
(a) Increased (b) Not changed
(c) Minimized (d) Stopped and they are killed

866. Which one of the following statements is correct?
(a) Vitamins are inorganic substances required by the body in small amounts for its normal functioning
(b) Vitamins are required in gram amounts each day, whereas trace elements are required in milligram amounts
(c) Carbohydrates, fats and proteins all have structural roles in the body
(d) Excessive intake of one mineral has no effect on the uptake of another

867. Which one of the following components is not a type of carbohydrate?
(a) Maltodextrins (b) Sterols (c) Lactose (d) Dietary fibre

868. Which one of the following statements is incorrect?
(a) Rickets is the major symptom of vitamin D deficiency
(b) Goitre is the major symptom of iodine deficiency
(c) Beriberi is the major symptom of vitamin B2 deficiency
(d) Scurvy is the major symptom of vitamin C deficiency

869. Which one of the following statements is correct?
(a) Proteins, fats and carbohydrates provide energy to the body, but alcohol does not
(b) Minerals and vitamins do not provide energy themselves and are not required to release energy from foods
(c) The metabolisable energy content of a food is greater than its heat of combustion
(d) Dietary fibre combusts completely in a bomb calorimeter but is indigestible to humans

870. The water activity (a_w) of milk powder is ---------------
(a) 0.10 (b) 0.20 (c) 0.30 (d) 0.40

871. Which of the following statements about the use of mass spectrometry in protein investigation is correct?
(a) Mass spectrometry involves ionized molecules in the liquid phase
(b) Mass spectrometry is used for analysing the three-dimensional shape of peptides
(c) Ionic fragments are separated according to their mass-to-charge ratio in Mass Spectrometry
(d) Mass spectrometry involves the separation of ionic fragments on a gel

872. In TLC plate's adsorbent is impregnated with ----------- to visualize the organic compounds under a UV lamp.
(a) Zinc sulfide (b) Amonium sulphate (c) EDTA (d) calcium chloride

873. Which of the following molecule(s) can be detected by ELISA?
(a) Proteins (b) Hormones (c) Antibodies (d) All of the above

874. In TLC experiment if distance travelled by compound is 1.7 cm from the base line and distance travelled by solvent is 5.0 cm then Rf value is--------------
(a) 0.34 (b) 0.24 (c) 2.94 (d) 0.54

875. -----------------is the solvent that carries the analyte.
(a) Eluent (b) Eluate (c) Solute (d) None of the above

876. A separation is which the mobile phase composition remains constant throughout the procedure is termed as--------------
(a) Isostatic elution (b) Gradient elution (c) Isocratic elution (d) None of the above

877. NIR spectra lies between the---------------
(a) Visible & mind infrared region (b) Visible and VU region
(c) Visible & X ray (d) None of the above

878. In infrared spectroscopy wave length is measured in wave numbers the unit of which is------
(a) in^{-1} (b) cm^{-1} (c) m^{-1} (d) None of the above

879. The Sephadex G-10 gel is used for separation of compounds having mol. weight-------------
(a) 5000-10000 (b) 5000-100000 (c) <700 (d) All of the above

880. The void volume of gel filtration column should be approximately---------------
(a) ¼ of the bed volume (b) ⅓ of the bed volume
(c) ½ of the bed volume (d) Equal of the bed volume

881. In ion exchange chromatography--------------compounds will elute first.
(a) Medium changed (b) Weakly changed
(c) Strongly changed (d) None of the above

882. What is shielding in NMR?
(a) Using a curved piece of metal to block an opponent's attack
(b) Putting metal around an Rf source
(c) When the magnetic moment of an atom blocks the full induced magnetic field from surrounding nuclei
(d) Blocking parts of a molecule from Rf radiation

883. ------------------ buffer is used for anion exchange chromatography.
(a) Formic acid (b) Tris (c) Malonic acid (d) Maleic acid

884. In reverse phase chromatography, the stationary phase is made-----------
(a) Non-polar (b) Polar
(c) Either non-polar or polar (d) None of the above

885. In TEM, the microscopic column is maintained under---------
(a) Magnetism (b) Pressure (c) Vacuum (d) None of the above

886. Mixture of bran, germ and shorts obtained during milling is called as --------------
(a) Flour (b) Mill feed (c) Semolina (d) Mill shorts

887. In which of the following process moisture content in the grain is adjusted?
(a) Conditioning (b) Parboiling (c) Milling (d) Parching

888. Mixture of bran and germ produced from milled wheat is known as---------------
(a) Mill feed (b) Wheat shorts (c) Flour (d) Semolina

889. Rheology is the study of -------------- properties of the dough.
(a) Elastic (b) Textural (c) Viscoelastic (d) Both a and b

890. ------------are important in determining gluten characteristics.
(a) Glutamine and gliadine (b) Gliadine and glutenine
(c) Aspargine and gliadine (d) Glutamine and glutenine

891. Wet milling tends to produce --------------- products.
(a) Coarser (b) Medium size (c) Finer (d) None of above

892. ------------ generally used as the anti staling agent in bread.
(a) GMS (b) GSM (c) MSG (d) SGM

893. Hydrolysis of triglycerides with alkali is called as-------------
(a) Titration (b) Hydrogenation (c) Saponification (d) Rancidity

894. Starch gets ----------- during extrusion cooking.
(a) Caramalize (b) Liquify (c) Gelatinize (d) Hydrolysed

895. Gliadine fraction of protein in gluten is responsible for ----------------
(a) Water holding capacity (b) Extensibility
(c) Elasticity (d) All of the above

896. For hydrogenation of oils, the catalyst required is-------------
(a) Iron (b) Aluminium (c) Magnesium (d) Nickel

897. Oils from cereal grains are rich in which vitamin?
(a) A (b) B (c) E (d) D

898. The paddy grain is mainly composed of --------------- starch granules.
(a) Hexagonal (b) Polygonal (c) Tringular (d) Circular

899. What are super foods?
(a) Foods full of nutrients (b) FDA-approved foods
(c) Foods that fight cancer (d) Organic foods

900. What do blueberries have over strawberries?
(a) More antioxidants (b) More vitamin C
(c) More folate (d) More fiber

901. The best way of cooking to maintain antioxidant activity in vegetables is by---------
(a) Steaming (b) Boiling (c) Microwaving (d) All of the above

902. Which of the following lacks antioxidant properties?
(a) Vit E (b) Omega 3 fatty acids (c) Vit K (d) Calcium

903. Among the following which are the antioxidants used in anti-aging products?
(a) Vitamins C and E (b) Coenzyme Q10
(c) Beta-carotene (d) All of the above

904. Which of the following food sources may contain probiotics?
(a) Yoghurt (b) Apple slices (c) Whole wheat bread (d) Canned green beans

905. Which of the following conditions would not benefit from the use of probiotics?
(a) Urinary tract infections in women (b) Lactose intolerance
(c) Irritable bowel syndrome (d) Upper respiratory infection

906. What is the composition of prebiotics?
(a) Fat (b) Fiber (c) Protein (d) Flavonoids

907. Probiotic cultures----------------
(a) Encourage the growth of beneficial bacteria in the gut
(b) Improve sight
(c) Enhance the body's capacity to absorb calcium
(d) None of the above

908. Plant sterols are------------
(a) A functional ingredient
(b) Classified as a probiotic substance
(c) Important in improving the health of the eye
(d) Able to stimulate the growth of bacteria in the intestine

909. In microwave heating, the source of microwave generator is ---------------
(a) Magnetron (b) Oven chamber (c) Control circuit (d) Transformer

910. In what way is a zoochemical different from a phytochemical?
(a) It has no beneficial compounds
(b) It is a beneficial compound that is derived from plants rather than animals
(c) It has more beneficial compounds
(d) It is a beneficial compound that is derived from animals rather than plants

911. Phytochemicals are------------
(a) Micronutrients (b) Macronutrients
(c) Found only in plants (d) Found only in animals

912. Which of the following statements is FALSE?
(a) Phytochemicals are nutrients necessary for sustaining health and life
(b) Phytochemicals protect plants from insects and microbes
(c) Phytochemicals have been shown to reduce the risk of cardiovascular disease and cataracts
(d) Phytochemicals are naturally occurring compounds found in plants

913. Which of the following statements is TRUE about the consumption of foods containing phytochemicals?
(a) They increase the risk of diseases of aging
(b) They reduce the risk of cancer
(c) They increase the rates of infection
(d) They increase the rates of blood clotting

914. Antioxidants neutralizes-----------
(a) Phytochemicals (b) Free radicals (c) Vitamins (d) All of the above

915. Lignans are part of which family of compounds?
(a) Carotenoids (b) Polyphenols (c) Phytosterols (d) None of the above

916. Which forms of flavonoids are also known as phyto-oestrogens?
(a) Flavanols (b) Flavones (c) Isoflavones (d) Anthocyanidins

917. A flavonoid glycone is one that has----------
(a) An amino acid attached (b) A fatty acid attached
(c) A sugar attached (d) None of the above

918. Catechin and epicatechin are--------
(a) Flavanols (b) Anthocyanidins (c) Phenolic acids (d) Carotenoids

919. Chlorogenic acid is a major polyphenol in----------
(a) Corn (b) Coffee (c) Soybeans (d) Vegetable oils

920. Which one of the following statements is correct?
(a) Carbohydrates are the body's major energy reserve
(b) When there is no further need for amino acids for growth or repair, they are broken down and used for energy
(c) The carbohydrates glucose, sucrose and fructose are non-glycaemic carbohydrates
(d) In addition to carbon, hydrogen and oxygen, all fats contain nitrogen

921. Which is the only non nutritive sweetener made from sucrose with unique properties?
(a) Sorbitol (b) Fructose (c) Sucralose (d) Stevia

922. -------------are essential fatty acids.
(a) Lipids (b) Omega three
(c) Linoleic and Linolenic (d) LDL and HDL

923. FSIS stands for------------
(a) Food Safety and Inspection Standards (b) Food Safety and Inspection Service
(c) Fiber Safety Inspection Service (d) Food Service and Inspection Safety

924. The HACCP process uses-------------to show the entire food processing operation.
(a) Personnel (b) Flow charts and diagrams
(c) Food processing software (d) Risk assessment

925. Swelling of the can is caused primarily by-----------

(a) Gas forming, anaerobic spore formers

(b) Gas forming, aerobic spore formers

(c) Both a and b

(d) None of the above

926. Pressure canners used in the commercial manufacture of canned products are known as------

(a) Steam blanchers (b) Retorts

(c) Plate exchangers (d) Sublimators

927. The purpose of using a leavening agent such as baking soda or baking powder in cakes and cookies is to provide a source of-------------

(a) Sodium dioxide (b) Carbon monoxide (c) Sodium monoxide (d) Carbon dioxide

928. An additive that can keep a compound, mixture or solution from changing its form or chemical nature is called a-------------

(a) Antioxidant (b) Buffer (c) Stabilizer (d) Preservative

929. Which of the following is for environment management?

(a) ISO-9000 (b) ISO-14000 (c) ISO-26000 (d) ISO-31000

930. Which European country is not a prime wine-making region?

(a) Spain (b) France (c) Belgium (d) Italy

931. Sparkling wines are named because they contain-----------

(a) Sulfur Dioxide (b) Carbon Dioxide (c) Hydrogen (d) Oxygen

932. The legislation that governs the sale of alcoholic beverages is called-----------

(a) Alcohol Control (b) Dram Shop

(c) Administrative Liquor Law (d) Consumption Statutes

933. Vermouths and aperitifs are---------------

(a) Sparkling wines (b) Fortified wines (c) Organic wines (d) Aromatic wines

934. Fortified wines contain about-----------

(a) 12% alcohol (b) 20% alcohol (c) 30% alcohol (d) 42% alcohol

935. Which of the following is not a component of good wine-making?

(a) Bottling (b) The soil

(c) The climate (d) The wine-making process

936. The juice that is extracted from the grapes is called-------------

(a) Clear run juice (b) Solution (c) Must (d) Wort

937. France is known for producing which of the following?

(a) Wine, champagne, and cognac (b) Wine, champagne, and beer

(c) Wine, port, and sherry (d) Wine, cognac, and port

938. A "vertical tasting" of wine involves------------

(a) Wines from the same year but from different vineyards or winemakers

(b) Wines from the same vineyards or winemakers but from different years

(c) Tasting wines while standing up

(d) None of the above

939. When at a restaurant and a small amount of wine is poured to taste, the purpose here is to---

(a) See if you like the wine (b) Make sure it is the wine you thought it was

(c) Make sure the wine is not spoiled (d) All of the above

940. What does "appellation" of wine refer to in France?

(a) Wines with an apple flavor (b) Place of origin of the wine

(c) Wines from the mountains (d) All of the above

941. Which does NOT necessarily mean a spoiled bottle of wine?

(a) Moldy Smell (b) Bottle Stink (c) Vinegar Smell (d) All of the above

942. Why red wines are usually aged in oak barrels?

(a) Stainless steel tanks would spoil the wine (b) To greatly improve the color

(c) Both a and b (d) To improve the wine's ability to age

943. If a wine is described as "hot," it most likely means it---------

(a) Has too high an alcohol content

(b) Is spicy flavored

(c) Is served at warmer than room temperature

(d) None of the above

944. How many basic tastes can the tongue perceive?

(a) 156 (b) 4 (c) 5 (d) 9

945. A magnetic tape which is of short length and is stacked onto surface of document is called ---------

(a) Inked stripe (b) Coded stripe (c) Magnetic stripe (d) Tagged stripe

946. A polished surface card that read by light and can store up to two million characters on it is best classified as------------

(a) Laser cards (b) Smart cards (c) Coded cards (d) Magnetic cards

947. Highly branched chains of glucose units result in------------

(a) Starch (b) Glycogen (c) Cellulose (d) Galactose

948. Condensation of glucose molecules results in formation of--------------

(a) Starch (b) Cellulose (c) Glycogen (d) Glucagon

949. Rich sources of vitamin B are--------------

(a) Liver (b) Fresh liver oils

(c) Green leafy vegetables (d) Egg yolk

950. Fats act as------------

(a) Insulating material (b) A constituent of protoplasm
(c) Restricts water loss (d) All of the above

951. Minerals do not include------------

(a) Iodine (b) Calcium (c) Sodium (d) Iron

952. A general feeling of irritability and tiredness may be due to lack of---------------

(a) Proteins (b) Carbohydrates (c) Fats (d) Vitamins

953. Linkage of amino acids is called a--------------

(a) Condensation reaction (b) Oxidation reaction
(c) Hydrolysis reaction (d) Reduction reaction

954. Structural component of hemoglobin is------------

(a) Manganese (b) Zinc (c) Iron (d) Chorine

955. Term used to describe all of chemical reactions occurring inside any organism is-------------

(a) Condensation (b) Oxidation (c) Polymerization (d) Metabolism

956. Thyroxin is secreted from-------------

(a) Thiamine deficiency (b) Theological inheritance
(c) Thyroid glands (d) Muscles used for breathing

957. Most of animal fats are-----------------

(a) Saturated fats (b) Unsaturated fats
(c) Monounsaturated fats (d) Monosaturated fats

958. Symptoms of kwashiorkor include---------------

(a) Cracked and scaly skin (b) Loss of appetite
(c) Excess sweating (d) Itching

959. Stored energy found in most foods is called as--------------

(a) Potential kinetic energy (b) Potential physical energy
(c) Potential chemical energy (d) Adrenosine triphosphate

960. Starch is broken down to form glucose units when hydrolyzed by-----------

(a) Alkaline base (b) Acidic base (c) Neutral base (d) Salty base

961. Bulk of undigested matter is made up of-----------

(a) Glycogen (b) Cellulose (c) Galactose (d) Stearic acid

962. Energy balance refers to the---------------

(a) Equation of rate of oxidation
(b) Amount of starch in take
(c) Equilibrium between energy intake and energy expenditures
(d) Equilibrium of rate of reduction

963. Vegetables fats are------------
(a) Saturated fats (b) Unsaturated fats
(c) Monounsaturated fats (d) Disaturated fats

964. Test of proteins is known as-------------
(a) Iodine Test (b) Biuret test
(c) Ethanol emulsion test (d) Benedict Test

965. Breakage of hydrogen bonds in protein coils results in---------------
(a) Activation of enzymes (b) Denaturation of proteins
(c) Expedites digestive system (d) Inactivation of protein molecules

966. Peristalsis is enabled through-------------
(a) Minerals (b) Vitamins (c) Trace elements (d) Dietary fiber

967. Highest energy value is provided by---------------
(a) Minerals (b) Fats (c) Starch (d) Glucose

968. Water-soluble vitamins need to be supplied------------
(a) Daily (b) Weekly
(c) Fortnightly (d) Can easily be stored in body for later use

969. A common fat to cause heart attack is---------------
(a) Unsaturated fats (b) Cholesterol
(c) Monounsaturated fat (d) Bisaturated fat

970. Water acts as a reactant in-----------
(a) Polymerization (b) Condensation (c) Oxidation (d) Hydrolysis

971. By nature, vitamins are--------------
(a) Fiber (b) Minerals
(c) Organic compounds (d) Inorganic compounds

972. Fiber is important for proper functioning of-----------
(a) Liver (b) Large intestine (c) Kidneys (d) Small intestine

973. Energy reserves are normally stored as-------------
(a) Lactose (b) Glucose (c) Fat (d) All of the above

974. Saturated fatty acids chains are---------------
(a) Straight (b) Curved (c) Twisted (d) Compressed

975. Coiled chains of proteins are linked through weaker bonds of----------
(a) Carbon atoms (b) Hydrogen atoms (c) Oxygen atoms (d) All of the above

976. Fat acts as a solvent for-------------
(a) Protoplasm (b) Fat-soluble vitamins (c) Insulating materials (d) Mineral ions

977. Sulfur and phosphorus are often present in-----------
(a) Proteins (b) Fats (c) Carbohydrates (d) Glucose

978. Glucose, fructose and galactose are------------

(a) Simple sugars (b) Polymer sugar (c) Complex sugar (d) Compound sugar

979. Water-soluble vitamins are--------------

(a) Stored in body (b) Excreted by body

(c) Released by body muscles (d) Released by bone-marrow

980. Water insolubility of glycogen makes it excellent for----------

(a) Polymerization (b) Storage materials (c) Late digestion (d) Weight loss

981. Obesity results if-------------

(a) Energy intake is lesser than energy expenditure

(b) Number of enzymes become more than food

(c) Energy expenditure is less than energy intake

(d) Energy expenditure is more than energy intake

982. Lack of vitamin B may result in-------------

(a) Poor eye sight (b) Dull skin (c) Tooth decay (d) Poor cellular respiration

983. A very high protein intake is------------

(a) Good for health (b) Good for kidneys and liver

(c) Dangerous for kidneys and liver (d) Helps in weight loss

984. In hydrolysis, larger molecules are broken down into smaller molecules by addition of-------

(a) Amino acids (b) Water (c) Grape sugar (d) Sucrose

985. A denatured enzyme---------------

(a) Looses its concave structure

(b) Looses its active sites

(c) Gets divided into many smaller enzymes

(d) Becomes even more quicker in chemical reactions

986. Nature of most intestinal enzymes is------------

(a) Acidic (b) Alkaline (c) Water based (d) Mineral based

987. Most of co-enzymes are------------

(a) Protein based (b) Water based (c) Easily denatured (d) Non-protein based

988. The action of enzyme depends upon------------

(a) Active site (b) Available site (c) Recognizable site (d) Detachable site

989. Substances on which enzymes act are called as--------------

(a) Prostrate (b) Hydrate (c) Glucate (d) Substrate

990. Structure of enzyme is-----------

(a) Two dimensional coiled strands (b) Three dimensional coiled strands

(c) Long unbranched chains (d) Short branched chains

991. During weight loss, an increased protein diet------------

(a) Is good for health (b) Helps build muscles and cuts fats

(c) Is necessary for vital processes (d) May increase amount of nitrogen in body

992. Vitamins----------

(a) Do not provide energy (b) Are inorganic in nature

(c) Cannot be hydrolyzed (d) Need special enzymes to be digested

993. Amino acid chain in a protein molecule is-------------

(a) Straight (b) Coiled (c) Curved (d) Concave

994. Bread and cereal group is rich in--------------

(a) Minerals (b) Carbohydrates (c) Fats (d) Vitamins

995. Dietary imbalances cause-------------

(a) Malnutrition (b) Starvation

(c) Eating only proteins (d) High cholesterol

996. As compared to carbohydrates, fats are lower in proportion of-----------

(a) Hydrogen (b) Oxygen (c) Carbon (d) All of the above

997. A damage to nervous system and limb paralysis is caused due to-------------

(a) Rickets (b) Beriberi (c) Diabetes (d) Scurvy

998. Cereals, potatoes and tapioca are common sources of------------

(a) Simple sugars (b) Complex sugars (c) Polysaccharides (d) Monosaccharides

999. Loss of equilibrium in energy intake and energy expenditure results in--------------

(a) Malnutrition (b) Over-nutrition (c) Kwashiorker (d) Both a and b

1000. Rickets is characterized by------------

(a) Knocked knees (b) Bowed legs (c) Swollen joints (d) Both a and b

1001. Process of condensing many small molecules to form one large molecule is called----------

(a) Polymerization (b) Condensation (c) Hydrolysis (d) Oxidation

1002. Glycogen is-----------

(a) Soluble in water (b) A form of fats

(c) Insoluble in water (d) A form of protein

1003. Cellulose is a form of------------

(a) Monosaccharide (b) Disaccharides

(c) Polymer (d) Polysaccharides

1004. In order to digest proteins, it is converted into------------

(a) Glucose (b) Fiber (c) Polypeptides (d) Casein

1005. According to healthy diet pyramid, fat, salt and sugar shall be---------

(a) Eaten more often

(b) Eliminated to avoid cholesterol, blood pressure and diabetes

(c) Eaten only sometimes

(d) Eaten least

1006. Enzymes degrade, alter or synthesize a food component through-------------
(a) Oxidation/reduction/isomerization (b) Hydrolysis/synthesis
(c) Group transfer (d) All of the above

1007. Sulphydryl oxidase is used for-------------
(a) Discoloration (b) Clarification of images
(c) UHT milk off flavour removal (d) All of the above

1008. Sugar results after hydrolysis of---------------
(a) Peptones (b) Carbohydrates (c) Trace elements (d) Fats

1009. Liver and muscles are common places of storage of-----------
(a) Glycogen (b) Amino acids (c) Lipids (d) Galactose

1010. In order to digest polypeptides, they are converted into--------------
(a) Glucose (b) Fiber (c) Amino acids (d) Polypeptides

1011. Hydrolysis cannot be carried out without-------------
(a) Water (b) Amino acids (c) Peptides (d) Mitochondria

1012. Unsaturated fatty acid chains bend at places of-----------
(a) Carbon molecules (b) Hydrogen molecules
(c) Oxygen molecules (d) All of the above

1013. A peptide bond is formed between----------
(a) Two starch molecules (b) Two amino acids
(c) Two galactose molecules (d) Two fat molecules

1014. In case of malnutrition, basic body needs are met through consumption of stored-----------
(a) Fats and glucose (b) Fats and glycogen
(c) Starch and cellulose (d) Peptones

1015. Binding of enzyme and substrate on active site is explained through------------
(a) Hide and seek hypothesis (b) Lock and hide hypothesis
(c) Lock and key hypothesis (d) Lock and run hypothesis

1016. Which is not an advantage of the fermented food?
(a) Makes the food more digestible (b) Increase storage life
(c) Synthesize vitamins (d) Decrease intestinal microflora

1017. In anaerobic (without O_2) environment, sugar ($C_nH_{2m}O_m$) in dough is converted into-------
(a) Glucose (b) Water (c) Alcohol (d) Carbon monoxide

1018. Fermentation which is carried by bacteria is called----------
(a) Lactic acid fermentation (b) Alcoholic fermentation
(c) Pyruvic fermentation (d) Acrylic fermentation

1019. Considering fermentation at industrial level, microorganism *Bacillus* is used to form-------
(a) Ethanol (b) Formic acid (c) Acrylic acid (d) Glycerol

1020. Despite keeping a check on weight, an increase in weight may be due to more------------
(a) Sodium in diet (b) Potassium in diet (c) Chlorine in diet (d) Manganese in diet

1021. Function of myoglobin is to------------
(a) Prevent swollen joints (b) Aid in muscular contraction
(c) Aid in blood clotting (d) Keeps retina intact

1022. ---------takes a liquid stream and separates the solute or suspension as a solid and the solvent into a vapour.
(a) Spray dryer (b) Freeze dryer (c) Drum dryerd (d) Pulse combustion dryer

1023. Softening of bones results in-------------
(a) Kwashiorkor (b) Rickets (c) Osteomalacia (d) Arthritis

1024. Liver produces urea to remove------------
(a) Excess hydrogen (b) Excess hydrogen peroxide
(c) Excess nitrogen (d) Excess starch

1025. Calcification of kidneys is caused due to-----------
(a) Lack of dietary fiber (b) Excess of sugar
(c) Excess of starch (d) Excess of fat-soluble vitamins

1026. Diseases like rickets, scurvy and beriberi occur due to deficiency of-----------
(a) Amino acids (b) Carbohydrates (c) Lipids (d) Vitamins

1027. Which of the following is not a product of fermentation?
(a) Lactate (b) Oxygen (c) Carbon dioxide (d) Ethanol

1028. Production of --------------occurs in the fermentation of grains.
(a) Must (b) Brine (c) Lactic acid (d) Carbon dioxide

1029. Applications of fermentation includes----------
(a) Cereal products (b) Dairy products (c) Beverage products (d) All of the above

1030. Common example of fermented beverage product is-------------
(a) Pickles (b) Beer (c) Bread and buns (d) Cheese and yogurt

1031. Which drying process is often used as an encapsulation technique in food and other industries?
(a) Rotary vacuum (b) Drum (c) Freeze (d) Spray

1032. Beyond optimum temperature, enzyme-----------
(a) Becomes flat (b) Becomes hyperactive
(c) Divides to form smaller enzymes (d) Becomes denatured

1033. Which is not fruit or vegetable based fermented product ?
(a) Wine (b) Beer (c) Vinegar (d) Sauerkraut

1034. Function of streptococcus during milk fermentation is to------------
(a) Produce aerobic environment (b) Produce anaerobic environment
(c) Form lactic acid (d) All of the above

1035. Pickles and sauerkraut share a common inoculum, which is------------
(a) *Lactobacillus plantarum* (b) *Lactobacillis bulgaricus*
(c) *Lactobacillis acidophilus* (d) *Saccharomyces cerevisiae*

1036. Which of the following waves have longest wave length----------
(a) X-rays (b) Microwaves (c) Gamma (d) Radio waves

1037. Catalytic hydrogenation of vegetable oils convert them in to---------
(a) Vegetable ghee (b) Free fatty acid (c) Trans fat (d) All of the above

1038. In dough, starch is digested into sugar through-----------
(a) Amylase (b) Protease (c) Maltase (d) Lactase

1039. The production of substances in industrial microbiology occur in the sequence of-----------
(a) Fermentation, downstream processing, removal of waste, inoculation
(b) Inoculation, downstream processing, fermentation, removal of waste
(c) Inoculation, fermentation, downstream processing, removal of waste
(d) Removal of waste, inoculation, fermentation, downstream processing

1040. Source of butyric acid is-----------
(a) Butter (b) Palm oil (c) Beef fat (d) Olive oil

1041. Greater number of carbon atoms in chain results in higher---------
(a) Melting point (b) Boiling point (c) Freezing point (d) Stability

1042. Derivatives of phosphatidic acid, which are composed of glycerol, fatty acids and phosphoric acids are known as-----------
(a) Phospholipids (b) Acylghycerols (c) Triglycerides (d) Esters

1043. High performance liquid chromatography (HPLC. cannot be used to-------------
(a) Identify the various pigments from a leaf extract
(b) Separate organic pesticides
(c) Determine the caffeine content in coffee samples
(d) Determine the mercury content in a fish sample

1044. Which of the following is stimulating beverage?
(a) Coffee (b) Soft drink (c) Pure juice (d) Milk

1045. Freeze-drying is a method for preserving materials, which are-------------
(a) Unstable in solution (b) Stable in air
(c) Unstable in air (d) Stable in solution

1046. The micro-organism which is present in both fresh and frozen juices is--------------
(a) *E. coli* (b) *Entereobacter aerogenes*
(c) *P. chrysogenum* (d) None of the above

1047. Layer of fat under mammalian skin acts as an----------
(a) Electric insulator (b) Thermal insulator
(c) Protection from germs (d) Protection from viruses

1048. Primary steps in protein purification includes--------------

(a) Homogenization (b) Differential centrifugation
(c) Solubilisation (d) All of the above

1049. In gas chromatography, the basis for separation of the components is the difference in ----

(a) Partition coefficients (b) Conductivity
(c) Molecular weight (d) Molarity

1050. The heat conduction in dry air is ----------

(a) Less rapid than in steam (b) More rapid than in steam
(c) Similar to steam (d) None of the above

1051. ---------------- offers superior mechanical strength.

(a) Glass (b) Metal container (c) Paper (d) Plastic

1052. In fractional distillation, a mixture of liquids is separated based on their -------------

(a) Boiling point (b) Solubility (c) Density (d) Chemical composition

1053. Membrane bioreactor is used for the production of --------------

(a) Amino acid (b) Antibiotics (c) Vitamins (d) All of the above

1054. The destruction of microorganisms by steam may be a --------------

(a) First order chemical reaction (b) Zero order chemical reaction
(c) Second order chemical reaction (d) None of the above

1055. Oil and water may be separated by using -------------

(a) A filter funnel (b) Chromatography paper
(c) A separating funnel (d) Condenser

1056. ------------numbers of phases are involved in freeze drying process.

(a) Three (b) One (c) Two (d) Four

1057. Pervaporation method involves---------------

(a) Removal of ions (b) Production of potable water
(c) Purification of aqueous streams (d) Separation and concentration of liquid mixture

1058. Under what conditions food poisoning bacteria may grow and produce toxins in vegetables?

(a) When thawed vegetables are held at refrigerated temperature for any considerable period
(b) When thawed vegetables are held at room temperature for any considerable period
(c) When thawed vegetables are held below refrigerated temperature for any considerable period
(d) None of the above

1059. A substance which can alter or speed up chemical reaction without itself being chemically unchanged is-------------

(a) Saliva (b) Pathogen (c) Catalyst (d) Influenza

1060. Galactose is a components of-------------

(a) Glucose (b) Lactose (c) Sucrose (d) Fructose

1061. Galactose and lactose are commonly found in-------------
(a) Animals other than mammals (b) Mammals
(c) Microscopic organisms (d) Plants
1062. Basal metabolism refers to amount of energy required-----------
(a) To eat dinner (b) To carry out all vital processes
(c) For running a marathon (d) To keep heart beating
1063. On boiling with benedict solution, reducing sugar will form a precipitate of------------
(a) Green color (b) Red color (c) Blue-black color (d) Violet color
1064. The "flash method" or "high temperature short time" method exposes fluids to------------
(a) Heat below 100°C (b) Heat at 62.3°C for 30 minutes
(c) Heat at 72°C for 15 seconds (d) Heat at 134°C for 1 second
1065. Salting, as a preservative-----------
(a) Plasmolyzes bacteria and fungi (b) Retards growth of *Staphylococcus aureus*
(c) Is used to prevent growth of halophiles (d) All of the above
1066. For black tea processing, fermentation of rolled leaves is done for-----------
(a) 0.5-1h (b) 2-5h (c) 8-10h (d) 12-15h
1067. Beverages are consumed for-------------
(a) Food value (b) Thirst quenching (c) Stimulating effect (d) All of the above
1068. Which of the following dryers is used for making baby food?
(a) Pneumatic conveyor dryer (b) Drum dryer
(c) Freeze dryer (d) Spray dryer
1069. Milk fermentation to produce cheese is done initially by inoculating with----------
(a) *Saccharomyces cerevisiae*
(b) *Streptococcus lactis and Lactobacillus* spp.
(c) *Acetobacter and Gluconobacter*
(d) *Lactobacillus bulgaricus and Streptococcus thermophilus*
1070. Rusting of iron is a -------------- process.
(a) Oxidation (b) Reduction (c) Absorption (d) Adsorption
1071. The process for converting sugar into ethanol is called as -------------
(a) Distillation (b) Desalination (c) Combustion (d) Fermentation
1072. For developing a new gluten-free pasta/macaroni product. Which ingredient would be suitable for use in the product?
(a) Rye flour (b) Rice flour (c) Semolina flour (d) Whole wheat flour
1073. Fractional distillation cannot be used to ----------
(a) Separate the components of liquid air
(b) Refine crude oil

(c) Test purity of a substance
(d) Separate methanol and water

1074. -----------is applied to the frozen product to accelerate sublimation.
(a) Heat (b) Air (c) Water (d) Solvent

1075. The membrane configuration which is not used in microfiltration is-------------
(a) Plate and frame (b) Hollow fiber
(c) Spiral wound (d) Tubular

1076. The different ACC's between food categories reflect the---------------
(a) Expected level of contamination of the raw material
(b) Potential for microbial growth during storage
(c) Potential shelf life
(d) All of the above

1077. Plate count of bacteria in foods generally use the plating medium consisting of-------------
(a) Peptone, yeast extract, glucose, sodium chloride, agar and distilled water
(b) Yeast extract, glucose, sodium chloride, agar and distilled water
(c) Peptone, glucose, sodium chloride, agar and distilled water
(d) Peptone, yeast extract, glucose, sodium chloride and distilled water

1078. Yeast and mould count determination requires---------------
(a) Nutrient agar (b) Acidified potato glucose agar
(c) Macconkey agar (d) Violet Red Bile agar

1079. The aromatic volatile components of spices are called -------------
(a) Spice oil (b) Spice fat (c) Spice gel (d) Spice paste

1080. ------------- is an important solvent for wet and edible materials.
(a) Hexane (b) Acetone (c) Chloroform (d) All of the above

1081. Which of the following holds true for Pulsed Electric field?
(a) It has been successful in pasteurizing milk, yogurt, soup etc.
(b) If there are no air bubbles present or the if food has low electrical conductivity, PEF is non-applicable
(c) It's a continuous process. It cannot be applied for non-pump able solid food products.
(d) All of the above

1082. Which of the following is also called ring dryer?
(a) Pneumatic conveyor dryer (b) Drum dryer
(c) Freeze dryer (d) Spray dryer

1083. The predominant micro-organisms in frozen foods are-------------
(a) Bacteria (b) Micro-coccus (c) Yeast and moulds (d) None of the above

1084. Sugar-processed citrus fruits that includes the rind is called-------------
(a) Jelly (b) Marmalade (c) Jam (d) Punch

1085. Proper flavour and aroma of coffee is associated with--------------
(a) Drying coffee beans (b) Curing coffee beans
(c) Roasting (d) All of the above

1086. The sugar contributes to beverage------------
(a) Sweetness (b) Calories
(c) Body and mouth feel (d) All of the above

1087. Chocolate liquor is a--------------
(a) Beverage (b) Semi plastic food (c) Plant extract (d) None of the above

1088. Cold-pasteurised fruit juice is being produced in response to which of the following?
(a) Dietary concerns (b) Environmental issues
(c) A demand for convenience (d) A technological development

1089. Which of the following is the most likely cause of food spoilage of ice-cream?
(a) Bacteria (b) Enzymes (c) Poor handling (d) Inadequate packaging

1090. Which of the following separation processes is used in the manufacture of pulp free orange juice?
(a) Sieving (b) Filtration (c) Evaporation (d) Sublimation

1091. The following processes are used to manufacture a product heating- grinding- sublimation. What is the product most likely to be?
(a) Flour (b) Peanut butter (c) Instant coffee (d) Custard powder

1092. The predominant kind of bacteria causing spoilage in fish at chilling temperature is---------
(a) *Species of Pseudomonas* (b) *Micrococcus*
(c) *Bacillus* (d) None of the above

1093. Which of the following combination of processing and preservation techniques works best for smoked products?
(a) Salt and acidification
(b) Heat and solid content
(c) Heat, salt, acidification/minimal moisture content
(d) Heat, salt, dipping in brine/ minimal moisture content

1094. Advantage of turbo dryer over cabinet tray dryer is---------------
(a) It is more effective (b) The feed is uniformly heated at the edges
(c) None of the above (d) Both a and b

1095. Which of the following is NOT a method used for size reduction?
(a) Cutting (b) Impact (c) Burning (d) Shear

1096. Which of the following holds well as a reason for storage of food at low temperatures?
(a) Respiration rate decreases (b) Growth of microbes decreases
(c) Humidity is less (d) All of the above

1097. Sterilization can be carried out by----------
(a) Heat (b) Radiation (c) Chemical agents (d) All of the above

1098. For the turbulent flow, the dispersion coefficient is correlated as a function of------------

(a) Reynolds number (b) Sherwood number

(c) Schmidt number (d) Grashof number

1099. The destruction of microorganisms by steam may be described as------------

(a) First order chemical reaction (b) Zero order chemical reaction

(c) Second order chemical reaction (d) None of the above

1100. Which of the following is true about packaging food prior to freezing?

(a) Insulates the food (b) Controls dehydration

(c) Both a and b (d) None of the above

1101. A psychrophilic halophile would be a microbe that prefers-------------

(a) Cold temperatures and increased amounts of salt

(b) Warm temperatures and increased amounts of pressure

(c) Cold temperatures and the absence of oxygen

(d) Warm temperatures and increased amounts of acid

1102. What nutrient do you get from the leavening agents in our breads?

(a) Protein (b) Iron

(c) Complex carbohydrates (d) None of the above

1103. Name of fiber supplement is-----------

(a) Ispaghol husk (b) Cumin husk (c) Coriander husk (d) Mint husk

1104. Concentrate of fruits and vegetable juices------------

(a) Favor the growth of *A. Niger* and *A. Flavus* species

(b) Favour the growth of yeast and of acid and sugar tolerant Leuconostoc and Lactobacillus species

(c) Favor the growth of Saprophytic bacteria

(d) None of the above

1105. During ripening, fruits produces------------

(a) Ammonia gas (b) Nitrogen gas (c) Ethylene gas (d) Oxygen gas

1106. Yeast is most likely to grow in frozen fruits during------------

(a) Slow thawing (b) Refrigeration

(c) Ambient temperature (d) None of the above

1107. Stimulating effect of coffee is due to ------------

(a) Caffeol (b) Caffeone (c) Caffeine (d) Tannin

1108. Most suitable method of fruit juice concentration is----------------

(a) High speed high evaporation (b) Low temperature vacuum evaporation

(c) Freeze concentration (d) Reverse osmosis

1109. Most common germicidal for water treatment is-----------

(a) Nitrogen (b) Chlorine (c) Benzene (d) Nitrous oxide

1110. ---------------enzyme is present in malted beverages.
(a) Lipase (b) Protease (c) Amylase (d) Phenolase

1111. In most fruit juices the major portion of total soluble solids is consists of----------
(a) Salt (b) Sugar (c) Vitamin (d) Mineral

1112. What does the body's cells use glucose for?
(a) Defense against bacteria (b) Energy
(c) Defense against viruses (d) Growth

1113. Which of the following is a result of freezer burn?
(a) Irreversibly changes color, texture, flavor and nutritive values of frozen foods
(b) Roast beef turns light brown
(c) Dehydration of food items takes place
(d) All of the above

1114. -------------- is added to fruits prior to freezing commercially to protect their quality.
(a) Vitamin A (b) Ascorbic acid (c) Water (d) None of the above

1115. Which of the following change takes place during the ageing process in making of ice creams?
(a) Decrease in viscosity (b) Proteins change slightly
(c) Fats turn into liquids (d) All of the above

1116. Without which of the given steps, would ice cream be an inedible hard frozen mass?
(a) Freezing (b) Whipping with air
(c) Hardening of ice cream (d) All of the above

1117. What could be the possible reason for minimal use of MSG in preserved/ canned food items?
(a) Reduction in spore formation (b) Increase in spore formation
(c) Facilitating multiplication of bacteria (d) None of the above

1118. Which of the following is a reason for addition of spice oils to certain foods for the reason of preservation?
(a) Decrease enzymes (b) Coagulation of certain mechanisms
(c) Reduces heat resistance of spores (d) All of the above

1119. Which acid is pectin made from?
(a) Pectic acid (b) Polygalacturonic acid
(c) Both a and b (d) None of the above

1120. Thermophilic flat sour spoilage is due to --------------
(a) Storing conditions (b) Temperature
(c) Equipment or ingredients (d) All of the above

1121. It is possible to maintain conditions of temperature and pressure whereby the physical state of food substrate can be maintained at a critical point for successful removal of water. This is called---------
(a) Freeze rehydration (b) Freezing
(c) Freeze dehydration (d) None of the above

1122. The number of OH group in fats can be expressed as-----------
(a) Polenske value (b) Reichert-Meissil value
(c) Acetyl value (d) Iodine value

1123. Which of the following in a jelly/ jam is responsible for the formation of gel?
(a) Pectin (b) Acid (c) Water (d) Sugar

1124. Carbon atoms in fatty acid are ------------ than those of sugars.
(a) Less reduced (b) More reduced (c) Less oxidized (d) More oxidized.

1125. -------------- is the insoluble precursor of pectin which when boiled in acid solution such as jelly making is hydrolysed to soluble pectin.
(a) Pectic substances (b) Protopectin
(c) Methyl alcohol (d) Pectic acid

1126. Which of the following is not an industrial product made by the fungus *Aspergillus niger*?
(a) Galactosidase (b) Citric acid (c) Gluconic acid (d) Lysine

1127. The packaging which helps to extend shelf life and improves quality, safety and sensory properties of packed food is known as ---------------
(a) Intelligent packaging (b) Active packaging
(c) NMBP packaging (d) Microbial packaging

1128. The degree of unsaturation of lipid can be measured as------------
(a) Saponification value (b) Iodine value
(c) Polenske value (d) Reichert Meissil value

1129. Fungal amylases using stationary culture with wheat bran utilizes-----------
(a) *A. oryzae* (b) *A. niger* (c) *A. flavus* (d) *S. cerevisiae*

1130. Why high concentration of sugars is not employed in the fermentation medium?
(a) Calcium lactate is not produced
(b) Calcium lactate crystallizes from the medium and slows down the process of fermentation
(c) High sugar concentration itself crystallizes
(d) None of the above

1131. Which of the following is NOT an advantage of dehydration under controlled conditions over sun drying?
(a) Cost (b) Time of drying (c) Quality of drying (d) None of the above

1132. The destruction of ascorbic acid is accelerated due to which of the following reasons?
(a) Oxygen (b) Copper ions
(c) Ascorbic acid oxidase enzyme (d) All of the above

1133. Which of the following is NOT a method of controlling biological forces on a food supply?
(a) Packaging and chemical additives
(b) Reducing free water content
(c) Increasing osmotic pressure
(d) None of the above

1134. Combining two or more than two films into a single film is known as------------

(a) Kraft paper (b) Lamination (c) Restorable pouches (d) End folds wrapping

1135. As a commercial packaging material, which is the most significant property of glass?

(a) Chemical inertness (b) Thermal stability

(c) Compressive strength (d) Impact resistance

1136. PET is manufactured by ------------------ method.

(a) Reforming (b) Stretch blow molding

(c) Blow molding (d) Injection molding

1137. In direct extrusion process at higher temperature which of the following is used to avoid friction?

(a) Oil (b) Lubricants (c) Molten glasses (d) Wax

1138. Which of the following spore forming bacteria may lead to overcooked and degraded canned food?

(a) Mesophiles (b) Psychrophiles (c) Thermophiles (d) All of the above

1139. Which of the following is NOT a method of controlling enzymatic browning?

(a) Sulphuring of fruits (b) Vacuum dehydration

(c) In-package desiccation (d) None of the above

1140. ------------ treatment of potatoes for dehydration offers a means to control heat damage during damage and to control browning.

(a) Magnesium chloride (b) Sodium chloride

(c) Calcium chloride (d) All of the above

1141. Which of the following ingredients is added to a jelly?

(a) Pectin (b) Acid (c) Sugar (d) All of the above

1142. For the manufacturing process of standard pasteurized milk, lactose is added to whole milk and sugar to ---------------

(a) Control crystallization of sugar present (b) Enhance crystallization of sugar

(c) For preservation purpose (d) None of the above

1143. Which of the following is added in the making of peanut butter to prevent oil separation?

(a) Salt (b) Dextrose (c) Water (d) Lecithin

1144. Which of the following is the process of anaerobic degradation of proteinaceous materials?

(a) Respiration (b) Fermentation (c) Putrefaction (d) All of the above

1145. Which of the following is NOT an essential fatty acid?

(a) Linoleic acid (b) LInolenic acid (c) Arachidonic acid (d) None of the above

1146. The solubility of fatty acids in water--------------

(a) Increase with increase in chain length and fewer the double bonds

(b) Increase with increase in chain length and increase in double bonds

(c) Decrease with chain length and decrease with number of double bonds

(d) Decrease with chain length and increase with double bonds

1147. In naturally occurring unsaturated fatty acids, the double bonds are in------- conformation.
(a) Cis conformation
(b) Trans conformation
(c) A mixture of cis and trans conformation
(d) Cis and trans conformation alternatively

1148. Which of the following statement is true?
(a) Cholesterols are present in both plants and animals
(b) Sterol is the only type of cholesterol in plants
(c) Cholesterol is not found in plants
(d) Cholesterol is completely absent in liver cells of cold blood animals

1149. At room temperature an unsaturated fatty acid will have---------------
(a) Liquid consistency (b) Solid consistency
(c) Waxy consistency (d) Cannot be predicted

1150. Which of the following is an example for phytosterol?
(a) β-sitosterol (b) Campesterol (c) Brassicasterol (d) All of the above

1151. Number of milligrams of KOH required to neutralize fatty acid present in 1 g of fat is called as --------------
(a) Potassium value (b) Acid value
(c) Saponification value (d) Iodine value

1152. Which of the following is a requisite for a microorganism to be used in fermentation and pickling?
(a) Microorganisms must be able to grow on the substrate
(b) Organism must have the ability to maintain physiological constancy under growing conditions
(c) Desired chemical changes should take place in the required conditions
(d) All of the above

1153. A vinegar containing 5 grams of acetic acid per 100 cubic centimeters is referred to as -----
(a) 0.5 grain (b) 5 grain (c) 50 grain (d) 500 grain

1154. The rate of conversion of alcohol to acetic acid depends on -------------
(a) The activity of the organism
(b) The amount of alcohol present and the amount of surface exposed per unit of volume
(c) The temperature
(d) All of the above

1155. Excessive intake of polished rice causes deficiency of-------------
(a) Vitamin A (b) Vitamin B1 (c) Vitamin D (d) Vitamin K

1156. Generally fats with unsaturated fatty acids are ------------ at room temperature.
(a) Solid (b) Liquid
(c) Liquid in the presence of oxygen (d) Liquid in the absence of oxygen

1157. Which of the following statement is true?
(a) Oxidative rancidity is observed more frequently in animal fats than vegetable fats
(b) Oxidative rancidity is observed more frequently in vegetable fats than animal fats
(c) Plants fats ado not undergo oxidative rancidity
(d) Oxidative rancidity can be effectively checked by dehydrogenation of fatty acids

1158. Affinity chromatography is used for the analysis and isolation of---------------
(a) Insoluble starch substances
(b) Enzyme tyrosinase
(c) Antibodies bound with a covalently-attached antigen on cellulose column
(d) All of the above

1159. Greater the number of carbon atom in chain of fatty acid------------
(a) The boiling point will be higher (b) The boiling point will be lesser
(c) The melting point will be higher (d) The melting point will be lower

1160. Which of the following information is required prior to launching a new product?
(a) Product and raw material specifications
(b) Process development
(c) Plant location and operation
(d) All of the above

1161. In paper manufacturing -------------------in wood pulp is removed by washing.
(a) Fibers (b) Lignin (c) Moisture (d) None of the above

1162. Out of these, which bacteria are found in fresh/refrigerated meat?
(a) *Lactobacillus* (b) *Serratia* (c) *Staphylococcus* (d) *Moraxella*

1163. Units for radiation energy is---------
(a) Radura (b) Gray (Gy) (c) Percentage (d) MeV

1164. Which of the following is an example for derived lipids?
(a) Steroids (b) Terpenes (c) Carotenoids (d) All of the above

1165. The best means of preventing growth of undesirable organisms in fermented juice is to ----
(a) Add strong, unpasteurized vinegar after alcoholic fermentation is complete
(b) Add sugar
(c) Add yeast
(d) Neither of the mentioned

1166. Which of the following is a reason of smoking meat?
(a) Development of flavor and preservation
(b) Creation of new products
(c) Development of color and protection from oxidation
(d) All of the above

1167. Which of the following is true about extrusion process?
(a) Structure is homogeneous
(b) No time is lost in changing the shape
(c) Service life of extrusion tool is too high
(d) Its leading end is in good shape as compared to rolling

1168. Pectinase, an enzyme used for increasing yield and for clarifying juice is obtained by-------
(a) *A. oryzae* (b) *A. niger* (c) *A.flavus* (d) All of the above

1169. The commonly used media for submerged bacterial fermentation at commercial scale is----
(a) Beet molasses (b) Liver extract (c) Germinated seed (d) None of the above

1170. How many acts are repealed by Food Safcty and Standards Act, 2006?
(a) 4 (b) 5 (c) 7 (d) 8

1171. Which portion of wheat is rich in starch?
(a) Endosperm (b) Bran (c) Germ (d) None of the above

1172. A toxic element which may contaminate food is ------------
(a) Zinc (b) Copper (c) Lead (d) Calcium

1173. Freeze crack is seen when foods are frozen by-----------------
(a) Air freezing (b) Contact freezing (c) Immersion freezing (d) Cryogenic freezing

1174. Puffed products are dried to a moisture content less than--------------
(a) 12% (b) 8% (c) 4% (d) 16%

1175. Which of these is a non-essential amino acid?
(a) Valine (b) Gylcine (c) Lycine (d) Tryptophan

1176. The most common hydrocarbon in natural gas is-----------------
(a) Methane (b) Butane (c) Propane (d) Ethane

1177. Which of the following is NOT a step in black tea manufacture?
(a) Drying/Firing (b) Rolling (c) Withering (d) Lump formation

1178. After toasting and before refining, oil is --------------
(a) Used (b) Made into pellets by hammer mill and stored
(c) Washed (d) All of the above

1179. Which of the following has highest protein content?
(a) Oat (b) Ragi (c) Wheat (d) Bajra

1180. Green tea is ----------------
(a) Orthodox tea (b) Fermented tea (c) Unfermented tea (d) Semi-fermented tea

1181. Which of these standards are responsible for trade-mark or symbol of product quality by keeping image of consumer by the industry?
(a) Legal standards (b) Company standards
(c) Industry standards (d) Grade standards

1182. Which of the following is a major role for the industry?
(a) GMP compliance (b) Product recall system
(c) Costumer service (d) All of the above

1183. Which of the following are NOT key constraints of the food processing industry?
(a) Inadequate quality control
(b) High packaging cost
(c) Low demand
(d) Poor infrastructure as in no cold storage, warehouse etc

1184. Which of the following is NOT a step in the process involved in dry milling of maize?
(a) Degermination (b) Sifting
(c) Removal of moisture (d) Diluting

1185. Oolong Tea is -------------
(a) Mixture of green and black teas (b) Partial oxidation takes place
(c) The leaves are greenish brown (d) All of the above

1186. Which of the following is NOT a part of a wheat kernel?
(a) Bran (b) Endosperm (c) Germ (d) None of the above

1187. Most commonly occurring fatty acids in nature are-----------
(a) Even number of carbon atoms in an un-branched chain of 12 – 24 carbons
(b) Odd number of carbon atoms in an un-branched chain of 12 – 24 carbons
(c) Even number of carbon atoms in a branched chain of 12 – 24 carbons
(d) Odd number of carbon atoms in a branched chain of 12 – 24 carbons

1188. The best source of trans fatty acid in diet is-------------
(a) Vegetables (b) Ground nut (c) Dairy products (d) Fruits

1189. Increased dietary uptake of trans fatty acid causes blood level--------------
(a) Increase of HDL (b) Increase of LDL
(c) Increase of LDL and decrease of HDL (d) Increase of HDL and decrease of LDL

1190. The sterol which acts as the precursor of vitamin D2 (provitamin D2) is------------
(a) Colcalciferol (b) Lanosterol (c) Cholesterol (d) Ergosterol

1191. Which of the following macro-molecule can be most structurally diverse among living world?
(a) Carbohydrates (b) Proteins (c) Nucleic acids (d) Lipids

1192. The number of OH groups in fatty acids can be expressed as----------
(a) Polenske number (b) Reichert-Meissl number
(c) Acetyl number (d) Iodine number

1193. Which of the following is NOT an omega-6 fatty acid?
(a) Alpha-linolenic acid (b) Linoleic acid
(c) Eicosadienoic acid (d) Arachidonic acid

1194. Rancidity of lipids of lipid rich food stuffs is due to------------

(a) Hydrogenation of unsaturated fatty acids

(b) Reduction of fatty acids

(c) Oxidation of fatty acids

(d) Dehydrogenation of saturation fatty acids

1195. Which of the following statement is true about saponification value?

(a) The shorter is the chain length of fatty acids the higher is the saponification value

(b) The shorter is the chain length of fatty acids, the lower will be the saponification value

(c) The higher the chain saturation of fatty acid, the lower will be saponification value

(d) The lower is the saturation of fatty acid, the higher will be saponification valuc

1196. The bacteria most often involved in the spoilage of fish are------------

(a) Part of the natural flora of the external slime of fishes and their intestinal contents

(b) Part of the natural flora of the internal slime of fishes only

(c) Both a and b

(d) None of the above

1197. What is the major health concern regarding the misuse of vitamin C supplements?

(a) Overload on kidneys (b) Poor absorption

(c) Weight gain (d) Toxicity

1198. What is most likely to be prevented by thawing frozen food in the refrigerator?

(a) Freezer burn (b) Contamination

(c) Activation of enzymes (d) Increase in microbial load

1199. The outbreaks traced to foods from food processing plants are contributed due to which of the following factor?

(a) Inadequate refrigeration

(b) Preparing food items too far in advance

(c) Food items re-infected after final heat processing

(d) All of the above

1200. Which of the following is NOT covered under 'Finished Product Quality Assurance'?

(a) Finished product monitoring (b) Special finished product survey

(c) Factory visits (d) None of the above

1201. Which of the following is a step under process development?

(a) Establish tentative process specifications

(b) Organize necessary pilot plant facilities

(c) Produce pilot plant products and submit to taste panels

(d) All of the above

1202. Wheat flour with water becomes -------------
(a) Elastic and substance is called gluten (b) Weak and substance is called lipid
(c) Hard and unable to cook (d) Wet and difficult to cook

1203. Which of the following is covered under 'Raw Material Quality Assurance'?
(a) Specifications (b) Survey program (c) Service (d) All of the above

1204. Which of the following produces the electron beam in magnetic deflection mass spectrometer?
(a) Tungsten filament (b) Quartz rod
(c) Silica (d) Rhodium filament

1205. In SDS electrophoresis, proteins are separated on basis of-------------
(a) Charge (b) Mass (c) Both a and b (d) Structure

1206. Packaging material should be----------------- to retain desirable properties of food products.
(a) Permeable (b) Impermeable (c) Semi permeable (d) None of the above

1207. A diet high in cholesterol is most likely to lead to disease in which organ of the body?
(a) Heart (b) Kidney (c) Liver (d) Pancreas

1208. Which mineral requires supplementation in the diet due to its naturally low levels in the soil?
(a) Calcium (b) Iodine (c) Iron (d) Sodium

1209. Which class of additive is used to absorb moisture from the atmosphere to prevent food drying out?
(a) Anti-caking agents (b) Antioxidants
(c) Humectants (d) Mineral salts

1210. Which of the following is true about turmeric?
(a) Green rhizomes are boiled in water till a froth comes out and white fumes appear giving out a characteristic odour.
(b) The softness of the cooked rhizomes depends on the curing conditions and process.
(c) After drying, the turmeric hardens up
(d) All of the above

1211. Food processing in India is concentrated in which sector, maximum?
(a) Organized sector (b) Unorganized sector
(c) Small scale (d) None of the above

1212. In the seed cooking step of pre-press solvent extraction of oil--------------
(a) Reducing oil viscosity promotes oil collection
(b) Temperature is increased to inactive/denature the enzymes
(c) Sometimes, very high temperature affects meal protein
(d) All of the above

1213. Which of the following statements is not true about mass spectrometry?
(a) Impurities of masses different from the one being analyzed interferes with the result
(b) It has great sensitivity

(c) It is suitable for data storage

(d) It is suitable for library retrival

1214. Chromatography with solid stationary phase is called-------------

(a) Circle chromatography

(b) Square chromatography

(c) Solid chromatography

(d) Adsorption chromatography

1215. Which of the following combination of processing and preservation techniques works best for smoked products?

(a) Salt and acidification

(b) Heat and solid content

(c) Heat, salt, acidification/minimal moisture content

(d) Heat, salt, dipping in brine/ minimal moisture content

1216. Liquefaction of starch to dextrin is carried out by------------

(a) Amylases (b) Cellulases (c) Pectinases (d) All of the above

1217. The unit of kinematic viscosity is-------------

(a) gm/cm-sec^2 (b) dyne-sec/cm^2 (c) gm/cm^2-sec (d) cm^2/sec

1218. Which among these is a factor for increased consumption of processed food in India?

(a) Changing lifestyles (b) Food habits

(c) Organized food retail (d) All of the above

1219. Oil refining is a step--------------

(a) After oil extraction (b) Before oil extraction

(c) Between oil cooking and flaking (d) None of the above

1220. Which of the following tests do coffee beans have to undergo?

(a) Visual test (b) Aroma test (c) Tasting by a cupper (d) All of the above

1221. Coated meat products require-------------

(a) Breading (b) Pre-dusting (c) Battering (d) All of the above

1222. Which of the following is a disadvantage of food processing?

(a) Canning of food leads to loss of Vitamin C

(b) Processed food adds empty calories to food constituting junk

(c) Some chemicals make the human and animal cells grow rapidly which is unhealthy

(d) All of the above

1223. Oil/lemon/vinegar + spices applied to meat is called -----------

(a) Marinating (b) Emulsifying (c) Fermenting (d) Coating

1224. Which of the following is true about fruits and vegetable processing?

(a) They get spoil very fast and hence need to be consumed soon

(b) They have high moisture content and should be kept in a cold, dark place

(c) They're tender and hence get spoiled easily

(d) All of the above

1225. Certain bacteria are added to minced meat products. This activity is followed by dehydration. What is this activity called?

(a) Coating (b) Freezing (c) Curing (d) Fermentation

1226. Which of the following is not related to post harvest losses?

(a) Post harvest losses can be reduced by adding value to products

(b) Packaging, storage, transportation areas are where losses take place

(c) Farmers don't earn much after adding value to products

(d) Value can be added to products by converting raw form into a more processed/refined form

1227. Which of the following dryers cannot take feed in the form of pastes/slurries/sticky items?

(a) Drum dryer (b) Screw conveyor dryer

(c) Pneumatic dryer (d) Spray dryer

1228. Oil seed cakes are obtained by-----------

(a) Solvent extraction (b) Mechanical pressing

(c) Both a and b (d) None of the above

1229. Greases proof paper are used in packaging of ------------

(a) Fresh fruits and vegetables (b) Jam and jelly

(c) Baked foods (d) Frozen foods

1230. The rigid metal containers are sterilized by using ---------------

(a) Hot water (b) Super heated steam

(c) H_2O_2 (d) None of the above

1231. Chymosin hydrolyses the bond between---------------

(a) Alanine and glycine (b) Phenyl alanine and methonine

(c) Glutamic acid and alanine (d) Alanine and phenyl alanine

1232. Which of the following is recent technique used for packaging of milk that offers economy?

(a) Cans (b) Glass bottles

(c) Plastic bottles (d) Flexible plastic pouches

1233. Which of the following statements about Nuclear Magnetic Resonance (NMR) is correct?

(a) A disadvantage of NMR analysis is that it requires a large amount of protein

(b) NMR is used to determine the amino acid sequence of proteins

(c) NMR requires crystallisation of proteins

(d) The majority of known protein structures have been determined using NMR

1234. Chromatography is used to separate-------------

(a) Solution (b) Mixtures (c) Molecules (d) Atoms

1235. Which of the following statements about column chromatography is correct?
(a) Affinity chromatography involves the attachment to the column matrix of ionic groups known that bind to the wanted protein
(b) In reverse phase chromatography the wanted protein can be selectively eluted by solutions of different hydrophobicities or ionic strengths
(c) Ion-exchange chromatography separates proteins according to their size
(d) Gel-filtration chromatography separates proteins on their ability to bind to specific groups on the column matrix

1236. Which of the following is an important criterion for the packaging of a food material?
(a) Economy and convenience of the package
(b) Appearance of the package
(c) Protection of the food
(d) All of the above

1237. Which of the following is used for shredding meat in the food industry?
(a) Ball mill (b) Hammer mill (c) Knife cutter (d) All of the above

1238. Which of the following is the disadvantage of gas chromatography?
(a) It is not a good method
(b) It cannot be used for qualitative analysis
(c) It cannot be used for separation of volatile components
(d) It does not provide direct identification

1239. Pattern on paper in chromatography is called--------------
(a) Chroming (b) Chroma (c) Chromatograph (d) Chromatogram

1240. Which of the following is NOT the correct method to enhance storage?
(a) Reducing pressure (b) Increasing ethylene
(c) Decreasing respiration (d) None of the above

1241. Which of the following statements is true?
(a) Polyethylene films have a good heat-sealability
(b) Polyvinylidene chloride films are gas and water vapor impermeable
(c) Nylon films have high physical strength
(d) All of the above

1242. When crystallization process takes place for a long time, the size of the crystals is ----------
(a) Small (b) Large (c) No crystals formed (d) None of the above

1243. With the increase in temperature, the holding time of the dough for cooling needs to be-----
(a) Increased (b) Decreased (c) Kept constant (d) None of the above

1244. Relations between time and temperature to determine the stability of food products is obtained by using which of the following data?
(a) Measure of product quality (b) The number of days stored
(c) The storage temperature (d) All of the above

1245. Which of the following is useful in trouble shooting for physical systems hazard?
   (a) Developing and maintaining flow diagrams to cover all food processing and physical systems and environment
   (b) Identifying all physical system hazards to safety and integrity of products
   (c) Establishing systems of control for all hazards and maintaining records of the same
   (d) All of the above

1246. Lipid content in flour results---------------
   (a) Low flour lipid requires more mixing (b) More flour lipid requires more mixing
   (c) Low flour lipid requires less mixing (d) Lipid has no effect on dough formation

1247. Which of the following has highest fat content?
   (a) Rice (b) Bajra (c) Maize (d) Oat

1248. Out of these, which bacteria is found in processed/cured meat?
   (a) *Moraxella* (b) *Pseudomonas* (c) *Alcaligenes* (d) *Lactobacillus*

1249. The chromaplate or thin layer chromatography plate is made up of--------------
   (a) Glass (b) Wood (c) Fibre (d) Metal

1250. Agarose is composed of long unbranched chains of uncharged------------
   (a) Proteins (b) Vitamins (c) Carbohydrates (d) RNA

1251. In gel electrophoresis fragments are separated on basis of------------
   (a) Size (b) Charge (c) Both a and b (d) None of the above

1252. Which of the following is the application of ion exchange chromatography?
   (a) The softening of hard water
   (b) The demineralization of water
   (c) The separation and determination of anions
   (d) All of the above

1253. Liquid chromatography can be performed in which of the following ways?
   (a) Only in columns (b) Only on plane surfaces
   (c) Either in columns or on plane surfaces (d) Neither in columns nor on plane surface

1254. Proteins are separated in an SDS-PAGE experiment on the basis of their-----------
   (a) Relative content of positively charged residue only
   (b) Relative content of negatively charged residue only
   (c) Size
   (d) Relative content of positively and negatively charged residue

1255. What are the uses of partition chromatography?
   (a) Separation of amino acids
   (b) Analysis of closely related aliphatic alcohols
   (c) Separation of sugar derivatives
   (d) All of the above

1256. Out of these which is/are irradiated for sterilization purpose?

(a) Spices (b) Potatoes (c) Rice (d) All of the above

1257. Which of the following parts of a plant are spices NOT made from?

(a) Bark (b) Leaf (c) Root (d) Cell

1258. When meat is passed through a coarse grinder plate it is called -------------

(a) Chunking (b) Flaking

(c) Restructured meat product (d) Restructured meat product and chunking

1259. Which of the following is true about sound ultrasound?

(a) Generates mechanical energy to enhance chemical action on surfaces

(b) Scrubbing action loosens the dirt particles and cleans the food particle

(c) Both a and b

(d) None of the above

1260. Which of the following is an advantage of food processing?

(a) Availability of seasonal food throughout the year

(b) Removal of toxins and preserving food for longer

(c) Adds extra nutrients to some food items

(d) All of the above

1261. In solvent extraction, the oil and miscella are moved in a---------------

(a) Co- current fashion (b) Counter- current fashion

(c) Cross- current fashion (d) None of the above

1262. An oil press is called ----------------

(a) Expresser (b) Boiler (c) Expeller (d) None of the above

1263. When two films are laminated by using some adhesive is known as-------------

(a) Thermo lamination (b) Wet lamination

(c) Extrusion lamination (d) None of the above

1264. In mass spectrometer, the ion currents are measured using which of the following?

(a) Scintillation counter (b) Ion counter

(c) Electrometer tube (d) Electric fields

1265. Which of the following are considered to be the lowest form of electromagnetic radiation?

(a) IR radiation (b) Micro waves (c) UV radiation (d) Radio waves

1266. The principle on which thin layer chromatography is based is that the-----------------

(a) Different compounds are absorbed on an absorbent to different degrees

(b) Different compounds are absorbed on an absorbent to same degrees

(c) Different compounds are adsorbed on an adsorbent to different degrees

(d) Different compounds are absorbed on an absorbent to same degrees

1267. What is shielding in NMR?

(a) Using a curved piece of metal to block an opponent's attack

(b) Putting metal around an Rf source

(c) When the magnetic moment of an atom blocks the full induced magnetic field from surrounding nuclei

(d) Blocking parts of a molecule from Rf radiation

1268. Ion exchange chromatography is based on the-------------

(a) Electrostatic attraction (b) Electrical mobility of ionic species

(c) Adsorption chromatography (d) Partition chromatography

1269. HPLC methods include--------------

(a) Liquid/liquid (partition) chromatography

(b) Liquid/solid (adsorption) chromatography

(c) Ion exchange and size exclusion chromatography

(d) All of the above

1270. Lacquers cans are used for packaging of ---------------products.

(a) Bakery products (b) Coloured fruit

(c) Frozen fruits (d) Milk powder

1271. Which of the following statements about the use of mass spectrometry in protein investigation are correct?

(a) Mass spectrometry is used to rapidly determine the molecular weight of a protein

(b) Mass spectrometry is only used to determine the sequence of peptides and not proteins

(c) Mass spectrometry can be used to investigate post-translational modifications

(d) Mass spectrometry involves separating ionic fragments on a gel

1272. In disaccharides, two monosaccharides are linked together by---------------

(a) Ionic bond (b) Covalent bond

(c) Hydrogen bond (d) Glycosidic bond

1273. A technique which separates charged particles using electric field is-------------

(a) Hydrolysis (b) Electrophoresis (c) Protein synthesis (d) Protein denaturing

1274. Gases used in modified atmospheric packaging are------------

(a) O_2, SO_2, CO_2 (b) O_2, CO_2, N_2 (c) CO_2, N_2, NH_3 (d) SO_2, NH_3, CO_2

1275. Easy open end cans are widely used in-------------- packaging.

(a) Smart (b) Fruit pulp

(c) Dried powder (d) Beverages and fruit juice

1276. --------------gas is used in modified atmosphere packaging of potato wafers.

(a) Carbon dioxide (b) Oxygen (c) Nitrogen (d) Carbon monoxide

1277. ------------ containers are widely used for packing mineral water.

(a) PVC (b) PET (c) PP (d) PVA

1278. In chromatogram, the area under the peak can be used to determine which of the following?

(a) Components of the sample (b) Amount of component in the sample

(c) Column efficiency (d) Column resolution

1279. What hazards does HACCP address?

(a) It is used to guide businesses through the process of identifying food safety hazards.

(b) It highlights inaccuracies in organizations administrative processes.

(c) It shows food handlers which utensils to use when cooking meals.

(d) It focuses on potential fire hazards within the food preparation area.

1280. Most gut bacteria can be found in the-----------

(a) Stomach (b) Small intestines (c) Large intestines (d) Urinary tract

1281. Which of the following techniques would be most useful to identify and quantify the presence of a known impurity in a drug substance?

(a) NMR (b) MS (c) IR (d) HPLC

1282. Which of the following is most often used in the chromatograph in gas chromatograph MS?

(a) Curvette (b) Paper support (c) Capillary tube (d) Flask

1283. Which of the following problems occur when combining gas chromatography and mass spectroscopy?

(a) Difference in operating pressures

(b) Reduction in sensitivity

(c) Direct identification is not possible

(d) It does not permit direct introduction of the effluent

1284. The system for measurement of ion intensity in GS-MS system consists of which of the following?

(a) Band pass amplifier (b) Narrow band amplifier

(c) Wide band amplifier (d) Low pass amplifier

1285. Which of the following is not an advantage of mass spectrometer?

(a) Simple (b) Rugged (c) High sensitivity (d) Unlimited mass range

1286. NMR spectroscopy is used for determining structure in which of the following material?

(a) Radioactive materials (b) Insoluble chemical compounds

(c) Liquids (d) Gasesristic feature

1287. Proteins can be visualized directly in gels by------------

(a) Staining them with the dye (b) Using electron microscope only

(c) Measuring their molecular weight (d) None of the above

1288. Fatty acids which are straight chained are found in-----------

(a) Animals (b) Plants (c) Micro-organisms (d) Unicellular plants

1289. Which of the following is not an application of high performance liquid chromatography?
(a) Elimination of undesirable substances from blood
(b) Separation of pharmaceutical drugs
(c) Analysis of proteins, drugs and explosives
(d) Separation of lipids, fatty acids and steroids

1290. The subunit molecular weight as well as the number of subunits in the quaternary structure can be determined by------------
(a) SDS-PAGE electrophoresis (b) Gel filtration chromatography
(c) Combining information from a and b (d) Isoelectric focusing

1291. Which of the following is non-climacteric type of fruit?
(a) Pineapple (b) Litchi (c) Grape (d) All of the above

1292. SO_2 content in pure KMS is -----------
(a) 25.5% (b) 75.5% (c) 78.2% (d) 57.7%

1293. Which enzyme is responsible for converting protopectin into pectin?
(a) PME (b) Proto-pectinase (c) Poly Galucturonase (d) Pectinase

1294. The term "three quarterful or full three quarter" is used to denote fruit maturity in which crop in the following?
(a) Banana (b) Mango (c) Tomato (d) Pineapple

1295. In mass spectrometer, the sample that has to be analyzed is bombarded with which of the following?
(a) Protons (b) Electrons (c) Neutrons (d) Alpha particles

1296. Probiotics are received by consuming----------
(a) Fermented foods (b) Vegetables
(c) Fruits (d) Cereals

1297. In SDS-PAGE, the protein sample is first---------
(a) Treated with a reducing agent and then with anionic detergent followed by fractionation by electrophoresis
(b) Fractionated by electrophoresis then treated with an oxidizing agent followed by anionic detergent
(c) Treated with a oxidizing agent and then with anionic detergent followed by fractionation by electrophoresis
(d) None of the above

1298. Fruits and leaves have a natural protective coatings of fatty acids known as-------------
(a) Waxes (b) Oil (c) Cutin (d) Cell wall

1299. In which extrusion process the direction of flow of metal is in same direction as that of ram?
(a) Direct (b) Indirect (c) Impact (d) Hydrostatic

1300. Prebiotics are-------------

(a) Bad" types of bacteria
(b) Fiber that sustains healthy gut bacteria
(c) Supplements for kids
(d) Dangerous to health

1301. What is the first stage of putting together a HACCP system?

(a) Putting together a HACCP team.
(b) Making sure everyone has the correct forms that have to be filled in.
(c) Determining the product lines and distribution channels that should be included in the HACCP plan.
(d) Creating a flow diagram that gives a simple and clear outline of the steps involved in the food process of the company.

1302. The analysis and separation of industrial products like soap and synthetic detergents is done using -----------------

(a) Thin layer chromatography
(b) Gas chromatography
(c) Ion exchange chromatography
(d) Partition chromatography

1303. In a native PAGE, proteins are separated on the basis of-------------

(a) Net negative charge
(b) Net charge and size
(c) Net positive charges size
(d) Net positive charge

1304. Which of the following statements about the use of mass spectrometry in protein investigation is correct?

(a) Mass spectrometry involves ionized molecules in the liquid phase
(b) Mass spectrometry is used for analysing the three-dimensional shape of peptides
(c) Ionic fragments are separated according to their mass-to-charge ratio in Mass spectrometry
(d) Mass spectrometry involves the separation of ionic fragments on a gel

1305. Fatty acids are usually derived from-------------

(a) Triglycerides (b) Phospholipids (c) Both a and b (d) None of the above

1306. Which of the following steps takes place after injection of feed in column chromatography?

(a) Detection of components
(b) Separation in the column
(c) Elution from the column
(d) Collection of eluted component

1307. In genome southern blotting can be used to identify-------------

(a) Sequences
(b) Number of sequences
(c) DNA fragments
(d) RNA sequence

1308. In which type of chromatography, the stationary phase held in a narrow tube and the mobile phase is forced through it under pressure?

(a) Column chromatography
(b) Planar chromatography
(c) Liquid chromatography
(d) Gas chromatography

1309. Lipids which are composed of glycerol and fatty acids are called------------

(a) Acylglycerols (b) Waxes (c) Phospholipids (d) Glycolipids

1310. In a gel filtration column----------------

(a) Smaller proteins enter the beads more readily

(b) Large proteins elute first

(c) Both a and b

(d) Large proteins enter the beads more readily

1311. Which of the following is true about extrusion process?

(a) Structure is homogeneous

(b) No time is lost in changing the shape

(c) Service life of extrusion tool is too high

(d) Its leading end is in good shape as compared to rolling

1312. NMR is the study of absorption of ---------------- by nuclei in a magnetic field?

(a) Radioactive radiation (b) Radio frequency radiation

(c) Microwaves of NMR (d) All of the above

1313. In an SDS-PAGE-------------

(a) Proteins are denatured by the SDS

(b) Proteins have the same charge-to-mass ratio

(c) Smaller proteins migrate more rapidly through the gel

(d) All of the above

1314. Which of the following assays could not be performed by gas chromatography?

(a) Characterization of volatile oils

(b) Measurement of drugs and metabolites in biological fluids

(c) Characterization of raw materials for drug synthesis

(d) Analysis of intravenous sodium chloride infusion

1315. Which type of chromatography is used for the structural analysis?

(a) Column chromatography

(b) Paper chromatography

(c) Partition chromatography

(d) Affinity chromatography

1316. Mass spectrometer separates ions on the basis of which of the following?

(a) Mass (b) Charge (c) Molecular weight (d) Mass to charge ratio

1317. What are the benefits of implementing HACCP?

(a) It assists businesses that work within the food preparation and supply industry to identify and manage key controls over processes and thereby ensure safe food.

(b) It helps organizations to keep accurate administrative records relating to food production.

(c) It keeps workers involved in food production on their toes.

(d) It helps to increase the profits of a company involved in food preparation.

1318. Polyacrylamide gel is usually used for-------------
(a) Proteins (b) DNA (c) Both a and b (d) Vitamins

1319. Select the correct statement from the following.
(a) Paper chromatography is a type of partition chromatography
(b) A special quality paper is used in paper chromatography
(c) Chromatography paper contains water trapped in it, which acts as stationary phase
(d) All of the above

1320. Which of the following is the richest source of iron?
(a) Mango (b) Bael (c) Pomegranate (d) Dry karonda

1321. Chillies are rich source of------------
(a) Vitamin A (b) Vitamin C (c) Vitamin A and C (d) Vitamin E

1322. According to FPO, the maximum limit of SO2 allowed in squashes and cordials is-----------
(a) 350 ppm (b) 500 ppm (c) 1000 ppm (d) 600 ppm

1323. What sectors is HACCP applicable to?
(a) The motor industry and specifically the production line.
(b) The airline industry.
(c) Software development.
(d) It is suitable to be implemented by organizations directly or indirectly involved in various sectors of the food industry and related supply chain.

1324. What does Critical Control Point mean?
(a) The point when food handlers must start to make administrative records in the HACCP system.
(b) The point when steam starts to rise from food being cooke(d)
(c) When bacteria starts to grow uncontrollably on food.
(d) It is a point, step or procedure at which control can be applied to prevent or eliminate a food safety hazard or reduce it to an acceptable level.

1325. Which of the following separate the ions according to their mass-to-charge?
(a) Ion source (b) Detector (c) Magnetic sector (d) Electric sector

1326. Chemical composition of an acylglycerols is esters of fatty acids and-------------
(a) Bases (b) Alcohol (c) Water (d) Acids

1327. The quantitative analysis is done using---------------
(a) Ion exchange chromatography
(b) Thin layer chromatography
(c) Gas chromatography
(d) Liquid chromatography

1328. As a preservative, benzoic acid is most effective against-----------
(a) Mould (b) Yeast (c) Bacteria (d) Virus

1329. Which enzyme is responsible for converting pectin into pectic acid?
(a) Pectinase (b) Proto-pectinase
(c) Pectic Methyl Esterase (d) Poly Galucturonase

1330. Chromatography is a physical method that is used to separate and analyze ------------
(a) Simple mixtures (b) Complex mixtures
(c) Viscous mixtures (d) Metals

1331. Gas chromatography can be performed in which of the following ways?
(a) Only in columns (b) Only on plane surfaces
(c) Either in columns or on plane surfaces (d) Neither in columns nor on plane surfaces

1332. Which of the following is not done using column chromatography?
(a) The identification of unknown compounds
(b) The determination of homogeneity of chemical substances
(c) Separation of inorganic cations or complexes
(d) Separation of geometric isomers

1333. Best maturity indices of orange is-----------
(a) TSS (b) Sugar % (c) Acid % (d) Brix : acid ratio

1334. In agarose gel electrophoresis, DNA is moved towards the---------------
(a) Cathode (b) Anode (c) DNA doesn't move (d) None of the above

1335. Which is the precursor of ethylene?
(a) Tryptophane (b) Methionine (c) ABA (d) IAA

1336. Tomato fruits for canning should be harvested at which stage?
(a) Matured green (b) Red ripe stage
(c) Immature green stage (d) Half-ripe/pink stage

1337. Vegetables are subjected to drying after-------------
(a) Sulfuring (b) Sulphitation (c) Blanching (d) Canning

1338. What is the maturity index for avocado?
(a) Sugar content (b) Acid content (c) Oil content (d) TSS

1339. If proteins are separated according to their electrophoretic mobility then type of electrophoresis is ----------------
(a) SDS page (b) Free flow electrophoresis
(c) Electro focusing (d) Affinity electrophoresis

1340. Using chromatogram as detector in chromatography, a graph is obtained between ----------- and time.
(a) Quantity (b) Density (c) Concentration (d) Specific gravity

1341. The procedure for mass spectroscopy starts with which of the following processes?
(a) The sample is bombarded by electron beam
(b) The ions are separated by passing them into electric and magnetic field
(c) The sample is converted into gaseous state
(d) The ions are detected

1342. Thin layer chromatography is-------------

(a) Partition chromatography

(b) Electrical mobility of ionic species

(c) Adsorption chromatography

(d) None of the above

1343. Which of the following is not a component of mass spectrometer?

(a) Inlet system (b) Sweep generator (c) Ion transducer (d) Mass analyzer

1344. In onion pink colour is due to------------

(a) Anthocyanin (b) Quercitin (c) Xanthophyll (d) Carotene

1345. Mold inhibitor used in bread is ------------

(a) Sodium or calcium propionate (b) Sodium chloride

(c) Calcium carbonate (d) None of the above

1346. Nisin is used as------------- in food industry.

(a) Emulsifier (b) Stabilizer (c) Sweetener (d) Antimicrobial agent

1347. A good reason to increase flow rate in chromatography is -------------

(a) To minimize band broadening due to mobile phase mass transport

(b) To minimize band broadening due to mass transport in the stationary phase

(c) To minimize eddy diffusion

(d) To minimize band broadening due to longitudinal diffusion

1348. HACCP can not only prevent cases of food poisoning, it can also---------------

(a) Increase a company's profits.

(b) Make foot taste better.

(c) Help a company to comply with relevant food law regulations.

(d) Make food look more palatable.

1349. What is a HACCP Plan?

(a) A form that has to be filled in by all food handlers.

(b) A food hygiene rating scheme.

(c) A written document which is based upon the seven principles of HACCP, which clearly states the safety procedures to be followed to identify any hazards that must be avoided, removed or reduced.

(d) A system used in food hygiene auditing.

1350. What is the main difference between HPLC and UPLC?

(a) HPLC is reverse-phase whereas UPLC is normal-phase

(b) UPLC employs smaller stationary-phase particle size

(c) HPLC and UPLC employ different mobile phases

(d) HPLC and UPLC employ different detection methods e) HPLC operates at higher mobile phase pressures

1351. In chromatography, the stationary phase can be --------------- supported on a solid.
(a) Solid or liquid (b) Liquid or gas (c) Solid only (d) Liquid only

1352. The region of electromagnetic spectrum for nuclear magnetic resonance is----------
(a) Microwave (b) Radio frequency (c) Infrared (d) UV-rays

1353. Amino acids detected by spraying the plate with ninhydrin solution is an example of--------
(a) Column chromatography (b) Thin layer chromatography
(c) Paper chromatography (d) Liquid chromatography

1354. Germination affects nutritive value of legumes by--------------
(a) Increase in vitamin C content (b) Decrease in trypsin inhibitor activity
(c) Increase in enzyme activity (d) All of the above

1355. Mass spectrometers are used to determine which of the following?
(a) Composition in sample (b) Concentration of elements in sample
(c) Relative mass of atoms (d) Properties of sample

1356. In chromatogram, the position of peaks on the time axis can be used to determine which of the following?
(a) Components of the sample (b) Amount of component in the sample
(c) Column efficiency (d) Column resolution

1357. Which of the following is normally done to convert the sample into gaseous state?
(a) Sample is pressurized (b) Chemical reactions are made to occur
(c) Sample is heated (d) Sample is cooled

1358. Purification of enzymes and proteins is done using-------------
(a) Affinity chromatography (b) Liquid chromatography
(c) Column chromatography (d) Thin layer chromatography

1359. In chromatography, which of the following can the mobile phase be made of?
(a) Solid or liquid (b) Liquid or gas (c) Gas only (d) Liquid only

1360. Which of the following is the most sensitive of the spectral methods?
(a) Absorption spectroscopy (b) Mass spectroscopy
(c) Flame emission spectroscopy (d) Atomic emission spectroscopy

1361. Milk and curry left over can be turned into sour and spoiled at------------
(a) High temperature (b) Very low temperature
(c) Room temperature (d) Constant temperature

1362. What is a HACCP team?
(a) A team of highly trained chefs.
(b) A group of people who have the skills and knowledge needed to develop, implement and maintain a HACCP system.
(c) A team of government investigators.
(d) An office based team of administrative officials who specialise in food hygiene matters.

1363. Which of the following statements about SDS polyacrylamide gel electrophoresis is correct?
(a) Wanted proteins can be tested for their biological activity after separation by SDS polyacrylamide gel electrophoresis
(b) Proteins are solubilized but not denatured when separated by SDS polyacrylamide gel electrophoresis
(c) SDS polyacrylamide gel electrophoresis separates proteins on the basis of charge
(d) SDS polyacrylamide gel electrophoresis separates proteins on the basis of size

1364. When glucose molecules are joined by α 1-4 linkage, type of carbohydrate obtained is------
(a) Lactose (b) Starch (c) Glycogen (d) None of the above

1365. GC- MS has been developed for which of the following systems?
(a) Packed column (b) Open tubular column
(c) Capillary column (d) Porous layer column

1366. ISO 9001: 2008 Quality management systems deals with-------------
(a) Fundamentals and vocabulary (b) Guidelines for performance improvements
(c) Requirements for quality management (d) Customer satisfaction

1367. FSSAI is located in 5 regions with head office located at-------------
(a) Hyderabad (b) Mumbai (c) Bangalore (d) New Delhi

1368. Sugar and salt act as preservatives by -------------
(a) Killing micro-organisms directly
(b) Increasing the acid content of food
(c) Increasing the water content of food
(d) Binding water so it is not available for micro-organisms

1369. FSSAI has issued an advisory banning the use of which material for wrapping and packaging of food items?
(a) Newspaper (b) Plastic (c) Polythene (d) None of the above

1370. SPS under WTO stands for-------------
(a) Standards, Prevention and Specification
(b) Sanitary and Phytosanitary Standards
(c) Specifications for products and supplements
(d) Safety and prevention of Sickness

1371. Food safety and Standards act, 2006 passed by Indian parliament and notified on----------
(a) 23rd July 2006 (b) 23rd June 2006 (c) 23rd August 2006 (d) 23rd November 2006

1372. Coffee is adulterated with------------
(a) Saw dust (b) Chicory (c) Ghee (d) All of the above

1373. AGMARK act 1937 comes under-------------
(a) Dept. of Consumer Affairs, Govt. of India
(b) Dept. of Agriculture and Cooperation
(c) Dept. of Marketing and inspection
(d) Dept. of Legal Meterology

1374. What is an example of biological hazard?
(a) Salmonella (b) Dirt (c) Cleaners (d) Antibiotics

1375. What is HACCP system for?
(a) Physical, Chemical and Biological Hazards
(b) A systematic analysis of all steps and regular monitoring of the control points
(c) Identifying the CCPs, including their location, procedure and process
(d) Accurately monitoring food hygiene hazards

1376. ISO 19011:2011 Quality Management System deals with---------------
(a) Specifications with guidance for use
(b) Guidelines for performance improvements
(c) Customer satisfaction
(d) Guidelines for quality and or environmental management system auditing

1377. How many scientific panels have been constituted in Food Authority?
(a) 16 (b) 7 (c) 9 (d) 5

1378. In mass spectrometer, the ions are sorted out in which of the following ways?
(a) By accelerating them through electric field
(b) By accelerating them through magnetic field
(c) By accelerating them through electric and magnetic field
(d) By applying a high voltage

1379. Which of the following is not a type of ion detector used in mass spectrometers?
(a) Electron multiplier (b) Flame emission detector
(c) Faraday cup collector (d) Photographic plates

1380. Hersperidinase is used for-------------
(a) Juice clarification (b) Juice debittering
(c) Off flavour reduction (d) RNA reduction in fish

1381. Milk digestibility is improved by using--------------
(a) Rnase (b) Lactase (c) Amylase (d) None of the above

1382. Newton's law of viscosity relates----------
(a) Intensity of pressure and rate of angular deformation
(b) Shear stress and rate of angular deformation
(c) Shear stress, viscosity and temperature
(d) Viscosity and rate of angular deformation

1383. Which is true about rennet?
(a) It is a mixture of protease chymosin and pepsin
(b) It is a mixture of rennin and pepsin
(c) Both a and b
(d) None of the above

1384. Stress strain relationship for Newtonian fluid is----------------
(a) Parabolic (b) Hyperbolic (c) Linear (d) Inverse type

1385. Which medium is used for the production of Penicillin using immobilized cells-------------
(a) 1% peptone medium (b) Glucose medium
(c) Yeast extract medium (d) LB broth

1386. Which of the following mainly serve to convert starch into high fructose corn syrup (HFCS)?
(a) Amylase (b) Gluco-isomerase (c) Gluco-amylase (d) All of the above

1387. The reduction in off flavour of beer is practiced through-------------
(a) Hersperidinase (b) Rnase (c) Invertase (d) Diacetyl reductase

1388. An ideal fluid is--------------
(a) One which obeys Newton's law of viscosity
(b) Frictionless and incompressible
(c) Very viscous
(d) Frictionless and compressible

1389. Which of the following enzyme is responsible for causing vitamin B deficiency disease beriberi?
(a) Ascorbic acid oxidase (b) Thiaminase
(c) Lipoxygenase (d) None of the above

1390 Which of the following metallic ion is there in ascorbic acid oxidase?
(a) Mg (b) Fe (c) Cu (d) Mn

1391. Fluidized bed bioreactors provide higher mass transfer rates than packed bed bioreactors because---------------
(a) Mixing is higher in fluidized bed bioreactors
(b) Particles move with the fluid in a fluidized bed bioreactor
(c) Immobilized particles are smaller in the fluidized bed bioreactors
(d) All of the above

1392. The oxygen transfer rate in a bioreactor will increase if----------------
(a) Oil is added (b) Antifoam is added
(c) Detergent like molecules are added (d) Increase in the reactor temperature

1393. Water insoluble enzymes can be prepared by using multifunctional agents that are bifunctional in nature and have--------------
(a) Low molecular weight (b) High equivalent weight
(c) Low reactivity (d) High molecular weight

1394. An aerated bioreactor will increase oxygen transfer rates with the addition of detergents because it--------------
(a) Enhances bubble coalescence (b) Causes bubbles to expand
(c) Discourages bubble coalescence (d) Increases the surface tension of the liquid

1395. Citrus juice debittering can be carried out using------------

(a) Limoninase (b) Anthocyanase (c) Inulinase (d) None of the above

1396. The immobilized enzyme produced by micro encapsulation technique provides---------

(a) An extremely large surface area (b) Smaller surface area

(c) High amount of solvent (d) Relatively smaller surface area

1397. Lysozyme is naturally present in--------------

(a) Egg white (b) Bacteria (c) Milk (d) All of the above

1398. In sterilization process, spore of which of the following organism is considered as control?

(a) *Bacillus subtilis* (b) *Clostridium botulinum*

(c) *Bacillus stearothermophilus* (d) *Aspergillus niger*

1399. Juice clarification extraction is facilitated by using-------------

(a) Cellulases (b) Amylase (c) Inulinase (d) Lactase

1400. Pascal-second is the unit of--------------

(a) Pressure (b) Kinematic viscosity (c) Dynamic viscosity (d) Surface tension

1401. --------------are the low molecular weight proteins or protein complexes produced by bacteria that inhibit or kill other bacteria.

(a) Antibiotics (b) Bacteriocins (c) Drugs (d) None of the above

1402. *Dhokla* is prepared from -------------

(a) Rice and chickpea (b) Rice and lentil

(c) Rice and blackgram (d) Rice and cowpea

1403. The predominant microorganism responsible for souring and gas production, in *idli* is---

(a) *B. trispora* (b) *S. cerevisiae*

(c) *Leuconostoc mesenteroides* (d) All of the above

1404. The methyl bromide used in fumigation of fruits has strong activity against------------

(a) Fungi (b) Yeast (c) Bacteria (d) Viruses

1405. The major hazards to the health of the consumers of fermented sausages is/are------------

(a) Salmonellae (b) *Escherichia coli* (EHEC.

(c) Both a and b (d) None of the above

1406. High Pressure Processing is a -------------- preservation technology.

(a) Post-packaging (b) *Pre*-packaging (c) Both a and b (d) None of the above

1407. Total Quality Management emphasizes-------------

(a) The responsibility of the quality control staff to identify and solve all quality-related problems

(b) A commitment to quality that goes beyond internal company issues to suppliers and customers

(c) A system where strong managers are the only decision makers

(d) A process where mostly statisticians get involved

1408. Among the tools of TQM, the tool ordinarily used to aid in understanding the sequence of events through which a product travels is a------------

(a) Pareto chart (b) Flow chart (c) Check sheet (d) Taguchi map

1409. Which of the following is true about ISO 14000 certification?

(a) It is a prerequisite for ISO 9000 certification

(b) It indicates a higher level of adherence to standards than ISO 9000

(c) It is only sought by companies exporting their goods

(d) It deals with environmental management

1410. The five elements in the management process are-----------

(a) Plan, direct, update, lead, and supervise

(b) Accounting/finance, marketing, operations, and management

(c) Plan, lead, organize, manage and control

(d) Plan, organize, staff, lead, and control

1411. Water and alcohol are easily separated by distillation because of their differences .

(a) Boiling points (b) Densities (c) Melting points (d) Colours

1412. Coffee powders can be made from which of the following dryers?

(a) Pneumatic conveyor dryer (b) Drum dryer

(c) Freeze dryer (d) Spray dryer

1413. A large and important group of compound which can be made by repetition of simple isoprenoid units, is called-------------

(a) Terpenoids (b) Steroids (c) Carotenoids (d) Terpenses

1414. Sources of insoluble dietary fiber does not includes-----------

(a) Cereals and wheat bran (b) Vegetables and fruits

(c) Butter and margarine (d) Oats and beans

1415. Ion exchange chromatography is based on the -----------

(a) A Electrostatic attraction (b) Electrical mobility of ionic species

(c) Adsorption chromatography (d) Partition chromatography

1416. ---------takes a liquid stream and separates the solute or suspension as a solid and the solvent into a vapour.

(a) Spray dryer (b) Freeze dryer (c) Drum dryer (d) Pulse combustion dryer

1417. Which of these would be a critical limit?

(a) Washing vegetables before using them.

(b) Cooking chicken to reach a temperature of 165°F (74°C. for 15 seconds.

(c) Checking the use by date on canned ingredients.

(d) The temperature food is kept at in a fridge.

1418. The amino acid residues in protein molecules are exclusively-----------

(a) l stereoisomer's (b) d stereoisomer's

(c) l and d stereoisomer's (d) Enantiomers

1419. Which one of the following pair is not correctly matched?

(a) Anaemia-low iron intake (b) Marasmus-low calorie intake

(c) Kwashiorker-low protein intake (d) Scurvy-low vitamin B12 intake

1420. Wheat germ oil is richest source of-----------

(a) Vitamin A (b) Vitamin D (c) Vitamin E (d) Vitamin K

ANSWER KEY

1. d	2. a	3. c	4. b	5. a	6. b	7. d	8. b	9. d	10. a
11. b	12. d	13. c	14. c	15. b	16. d	17. b	18. d	19. c	20. b
21. a	22. c	23. c	24. a	25. a	26. c	27. a	28. c	29. d	30. c
31. b	32. d	33. b	34. b	35. d	36. c	37. d	38. d	39. a	40. a
41. b	42. b	43. c	44. d	45. c	46. d	47. c	48. b	49. b	50. c
51. c	52. d	53. c	54. b	55. c	56. d	57. b	58. d	59. c	60. d
61. a	62. a	63. b	64. b	65. a	66. a	67. d	68. b	69. a	70. b
71. c	72. d	73. b	74. b	75. b	76. b	77. c	78. a	79. a	80. b
81. a	82. c	83. c	84. d	85. a	86. b	87. a	88. a	89. d	90. d
91. c	92. d	93. c	94. b	95. d	96. b	97. d	98. c	99. a	100. d
101. d	102. a	103. b	104. d	105. d	106. c	107. a	108. d	109. c	110. b
111. d	112. a	113. b	114. d	115. d	116. d	117. d	118. c	119. a	120. d
121. b	122. c	123. c	124. c	125. d	126. d	127. a	128. d	129. a	130. d
131. d	132. c	133. b	134. a	135. a	136. c	137. c	138. d	139. b	140. d
141. c	142. c	143. b	144. a	145. a	146. a	147. d	148. a	149. d	150. a
151. b	152. c	153. d	154. a	155. c	156. c	157. a	158. a	159. b	160. d
161. d	162. a	163. b	164. a	165. d	166. b	167. a	168. c	169. b	170. c
171. b	172. d	173. b	174. a	175. d	176. c	177. d	178. c	179. d	180. d
181. a	182. d	183. a	184. b	185. a	186. d	187. d	188. b	189. a	190. d
191. d	192. b	193. c	194. b	195. d	196. d	197. b	198. d	199. b	200. c
201. c	202. c	203. c	204. a	205. c	206. b	207. c	208. c	209. d	210. b
211. a	212. d	213. b	214. d	215. b	216. c	217. a	218. b	219. d	220. d
221. d	222. a	223. b	224. c	225. a	226. a	227. b	228. d	229. c	230. a
231. b	232. a	233. c	234. b	235. d	236. c	237. b	238. c	239. d	240. c
241. c	242. a	243. c	244. c	245. b	246. b	247. a	248. c	249. b	250. a
251. a	252. c	253. b	254. d	255. c	256. a	257. b	258. c	259. a	260. d
261. d	262. a	263. c	264. c	265. d	266. a	267. c	268. a	269. d	270. c
271. b	272. a	273. d	274. c	275. d	276. b	277. b	278. d	279. b	280. c

281. d	282. d	283. c	284. d	285. d	286. b	287. d	288. d	289. a	290. d
291. d	292. b	293. d	294. c	295. d	296. b	297. d	298. a	299. d	300. b
301. a	302. c	303. b	304. b	305. a	306. b	307. a	308. b	309. d	310. a
311. b	312. d	313. d	314. d	315. d	316. a	317. d	318. d	319. d	320. b
321. d	322. c	323. d	324. d	325. d	326. d	327. d	328. c	329. a	330. b
331. a	332. a	333. b	334. b	335. b	336. d	337. d	338. a	339. b	340. c
341. c	342. d	343. c	344. d	345. c	346. d	347. d	348. b	349. c	350. d
351. c	352. c	353. b	354. c	355. d	356. b	357. b	358. c	359. c	360. a
361. b	362. c	363. a	364. c	365. d	366. a	367. c	368. d	369. b	370. a
371. d	372. a	373. d	374. b	375. c	376. d	377. a	378. c	379. b	380. c
381. b	382. a	383. d	384. c	385. d	386. b	387. d	388. d	389. c	390. c
391. a	392. d	393. d	394. d	395. b	396. c	397. d	398. a	399. c	400. b
401. d	402. b	403. a	404. b	405. b	406. a	407. b	408. c	409. b	410. d
411. c	412. d	413. b	414. d	415. c	416. d	417. d	418. d	419. c	420. b
421. a	422. c	423. d	424. a	425. b	426. a	427. c	428. b	429. c	430. b
431. b	432. c	433. d	434. c	435. c	436. c	437. b	438. b	439. d	440. d
441. b	442. a	443. d	444. a	445. b	446. b	447. c	448. a	449. d	450. c
451. a	452. d	453. d	454. a	455. c	456. d	457. b	458. a	459. c	460. a
461. b	462. d	463. c	464. b	465. b	466. d	467. c	468. a	469. c	470. a
471. a	472. c	473. c	474. d	475. d	476. a	477. c	478. d	479. d	480. d
481. c	482. b	483. d	484. a	485. b	486. a	487. a	488. b	489. a	490. b
491. c	492. b	493. b	494. d	495. a	496. d	497. c	498. b	499. b	500. c
501. b	502. d	503. d	504. b	505. a	506. d	507. c	508. b	509. d	510. d
511. b	512. d	513. c	514. b	515. b	516. b	517. b	518. d	519. c	520. b
521. a	522. b	523. d	524. d	525. c	526. b	527. a	528. c	529. d	530. c
531. d	532. c	533. d	534. b	535. a	536. c	537. d	538. a	539. b	540. d
541. c	542. d	543. a	544. d	545. c	546. b	547. b	548. b	549. b	550. c
551. a	552. d	553. c	554. d	555. c	556. c	557. b	558. c	559. b	560. b
561. d	562. d	563. b	564. c	565. b	566. a	567. b	568. d	569. a	570. c
571. c	572. b	573. d	574. d	575. b	576. d	577. a	578. d	579. a	580. d
581. a	582. a	583. d	584. a	585. d	586. c	587. d	588. c	589. d	590. d
591. b	592. b	593. a	594. d	595. d	596. a	597. d	598. d	599. d	600. a
601. b	602. c	603. b	604. d	605. b	606. a	607. b	608. a	609. b	610. d
611. c	612. a	613. a	614. b	615. c	616. a	617. a	618. b	619. a	620. b
621. c	622. d	623. d	624. a	625. c	626. d	627. c	628. a	629. c	630. c
631. d	632. c	633. d	634. a	635. a	636. a	637. a	638. d	639. b	640. b
641. b	642. a	643. c	644. d	645. d	646. b	647. d	648. b	649. d	650. d
651. a	652. b	653. b	654. b	655. d	656. a	657. d	658. d	659. d	660. c
661. c	662. b	663. b	664. c	665. d	666. d	667. b	668. a	669. c	670. d

671. d	672. b	673. d	674. a	675. c	676. b	677. c	678. d	679. d	680. c
681. d	682. d	683. c	684. b	685. a	686. a	687. a	688. c	689. d	690. a
691. d	692. d	693. c	694. b	695. d	696. b	697. d	698. b	699. a	700. b
701. d	702. b	703. c	704. c	705. b	706. b	707. a	708. d	709. a	710. c
711. d	712. b	713. d	714. d	715. d	716. d	717. b	718. d	719. a	720. c
721. b	722. c	723. d	724. d	725. a	726. d	727. d	728. d	729. c	730. d
731. d	732. b	733. a	734. c	735. c	736. b	737. d	738. d	739. b	740. d
741. c	742. d	743. a	744. c	745. c	746. a	747. b	748. d	749. b	750. b
751. b	752. d	753. a	754. c	755. a	756. d	757. d	758. d	759. c	760. c
761. c	762. b	763. d	764. d	765. d	766. b	767. b	768. d	769. d	770. c
771. a	772. b	773. a	774. a	775. a	776. a	777. a	778. b	779. b	780. b
781. d	782. b	783. d	784. b	785. b	786. d	787. a	788. b	789. a	790. c
791. c	792. a	793. a	794. c	795. c	796. d	797. b	798. a	799. d	800. b
801. c	802. c	803. a	804. a	805. c	806. a	807. c	808. a	809. d	810. d
811. d	812. a	813. d	814. a	815. b	816. d	817. c	818. c	819. a	820. c
821. d	822. a	823. c	824. d	825. d	826. a	827. a	828. d	829. d	830. d
831. c	832. b	833. d	834. d	835. d	836. c	837. d	838. d	839. d	840. b
841. c	842. d	843. d	844. d	845. b	846. d	847. a	848. d	849. c	850. b
851. c	852. d	853. a	854. a	855. a	856. c	857. b	858. b	859. d	860. c
861. c	862. a	863. d	864. d	865. c	866. c	867. b	868. c	869. d	870. b
871. c	872. a	873. d	874. a	875. a	876. c	877. a	878. b	879. c	880. b
881. b	882. c	883. b	884. a	885. c	886. b	887. a	888. b	889. c	890. b
891. c	892. a	893. c	894. c	895. b	896. d	897. c	898. b	899. a	900. a
901. a	902. d	903. d	904. a	905. d	906. b	907. a	908. a	909. a	910. d
911. c	912. a	913. b	914. b	915. b	916. c	917. c	918. a	919. b	920. b
921. c	922. c	923. b	924. b	925. a	926. b	927. d	928. c	929. b	930. c
931. b	932. d	933. d	934. b	935. a	936. c	937. a	938. b	939. c	940. b
941. b	942. d	943. a	944. b	945. c	946. a	947. b	948. d	949. a	950. d
951. a	952. d	953. a	954. c	955. d	956. c	957. a	958. a	959. c	960. b
961. b	962. c	963. b	964. b	965. b	966. d	967. b	968. a	969. b	970. d
971. c	972. b	973. c	974. a	975. b	976. b	977. a	978. a	979. b	980. b
981. c	982. d	983. c	984. b	985. b	986. b	987. d	988. a	989. d	990. b
991. d	992. a	993. b	994. b	995. a	996. b	997. b	998. c	999. c	1000. d
1001. a	1002. c	1003. d	1004. c	1005. d	1006. d	1007. c	1008. b	1009. a	1010. c
1011. a	1012. b	1013. b	1014. b	1015. c	1016. d	1017. c	1018. a	1019. c	1020. a
1021. b	1022. a	1023. c	1024. c	1025. d	1026. d	1027. b	1028. d	1029. d	1030. b
1031. d	1032. d	1033. c	1034. b	1035. a	1036. d	1037. a	1038. a	1039. c	1040. a
1041. a	1042. a	1043. d	1044. a	1045. a	1046. b	1047. b	1048. d	1049. a	1050. a

1051. b	1052. a	1053. d	1054. a	1055. c	1056. a	1057. d	1058. b	1059. c	1060. b
1061. b	1062. b	1063. b	1064. c	1065. a	1066. b	1067. d	1068. b	1069. b	1070. a
1071. d	1072. b	1073. c	1074. a	1075. d	1076. d	1077. a	1078. b	1079. a	1080. a
1081. d	1082. a	1083. c	1084. b	1085. c	1086. d	1087. b	1088. d	1089. c	1090. b
1091. c	1092. a	1093. d	1094. b	1095. c	1096. d	1097. d	1098. a	1099. a	1100. c
1101. a	1102. d	1103. a	1104. b	1105. c	1106. a	1107. c	1108. c	1109. b	1110. c
1111. b	1112. b	1113. d	1114. b	1115. b	1116. b	1117. a	1118. c	1119. b	1120. c
1121. c	1122. c	1123. a	1124. b	1125. b	1126. d	1127. b	1128. b	1129. a	1130. b
1131. a	1132. d	1133. a	1134. b	1135. a	1136. b	1137. c	1138. c	1139. d	1140. c
1141. d	1142. a	1143. b	1144. c	1145. c	1146. c	1147. a	1148. c	1149. a	1150. d
1151. b	1152. d	1153. c	1154. d	1155. b	1156. b	1157. a	1158. d	1159. c	1160. d
1161. b	1162. d	1163. b	1164. d	1165. a	1166. d	1167. b	1168. d	1169. a	1170. d
1171. a	1172. c	1173. d	1174. b	1175. b	1176. a	1177. d	1178. b	1179. a	1180. c
1181. b	1182. d	1183. c	1184. d	1185. d	1186. d	1187. a	1188. c	1189. c	1190. d
1191. d	1192. c	1193. a	1194. c	1195. a	1196. a	1197. a	1198. d	1199. d	1200. d
1201. d	1202. a	1203. d	1204. a	1205. b	1206. b	1207. a	1208. b	1209. c	1210. d
1211. b	1212. d	1213. a	1214. d	1215. d	1216. a	1217. d	1218. d	1219. a	1220. d
1221. d	1222. d	1223. a	1224. d	1225. d	1226. c	1227. c	1228. b	1229. c	1230. b
1231. b	1232. d	1233. a	1234. b	1235. b	1236. d	1237. c	1238. d	1239. d	1240. b
1241. d	1242. b	1243. a	1244. d	1245. d	1246. a	1247. d	1248. d	1249. a	1250. c
1251. c	1252. d	1253. c	1254. d	1255. d	1256. d	1257. d	1258. d	1259. c	1260. d
1261. b	1262. c	1263. b	1264. c	1265. d	1266. c	1267. c	1268. a	1269. d	1270. b
1271. a	1272. d	1273. b	1274. b	1275. d	1276. c	1277. b	1278. b	1279. a	1280. c
1281. d	1282. c	1283. a	1284. c	1285. c	1286. c	1287. a	1288. a	1289. a	1290. c
1291. d	1292. d	1293. b	1294. a	1295. b	1296. a	1297. a	1298. a	1299. a	1300. b
1301. a	1302. b	1303. b	1304. c	1305. c	1306. b	1307. b	1308. a	1309. a	1310. c
1311. b	1312. b	1313. d	1314. d	1315. b	1316. d	1317. a	1318. c	1319. d	1320. d
1321. c	1322. a	1323. d	1324. d	1325. c	1326. b	1327. c	1328. b	1329. c	1330. b
1331. a	1332. c	1333. d	1334. a	1335. b	1336. b	1337. c	1338. c	1339. a	1340. c
1341. c	1342. c	1343. b	1344. b	1345. a	1346. d	1347. d	1348. c	1349. c	1350. b
1351. a	1352. b	1353. c	1354. d	1355. c	1356. a	1357. c	1358. a	1359. b	1360. b
1361. c	1362. b	1363. d	1364. b	1365. a	1366. c	1367. d	1368. d	1369. a	1370. b
1371. c	1372. b	1373. a	1374. a	1375. b	1376. d	1377. c	1378. c	1379. b	1380. a
1381. b	1382. b	1383. c	1384. c	1385. b	1386. d	1387. d	1388. b	1389. b	1390. c
1391. b	1392. c	1393. a	1394. c	1395. a	1396. a	1397. d	1398. c	1399. a	1400. c
1401. b	1402. a	1403. c	1404. c	1405. c	1406. a	1407. b	1408. b	1409. d	1410. c
1411. a	1412. d	1413. a	1414. c	1415. a	1416. a	1417. b	1418. a	1419. d	1420. c